Cluster Computing
for **Robotics** and
Computer Vision

Cluster Computing
for **Robotics** and
Computer Vision

Damian M Lyons

Fordham University, USA

 **World Scientific**

NEW JERSEY · LONDON · SINGAPORE · BEIJING · SHANGHAI · HONG KONG · TAIPEI · CHENNAI

Published by

World Scientific Publishing Co. Pte. Ltd.

5 Toh Tuck Link, Singapore 596224

USA office: 27 Warren Street, Suite 401-402, Hackensack, NJ 07601

UK office: 57 Shelton Street, Covent Garden, London WC2H 9HE

British Library Cataloguing-in-Publication Data
A catalogue record for this book is available from the British Library.

CLUSTER COMPUTING FOR ROBOTICS AND COMPUTER VISION

ISBN-13 978-981-283-635-9
ISBN-10 981-283-635-7

Typeset by Stallion Press
Email: enquiries@stallionpress.com

Printed in Singapore by World Scientific Printers.

To Jeanie Shippey Lyons
and Jeani Shannon Lyons

Preface

Robotic technology is expanding into many new fields including the military, professional service, home/consumer and educational fields. However, meeting the computational demands of these new applications requires harnessing the latest computational technologies. Cluster computing is an approach to providing supercomputer performance from a collection of off-the-shelf computer systems. With the advent of consumer multi-core processors even single computer systems can furnish several processor cores for a cluster.

The primary objective of this text is to give professionals and students working in the cluster computing field, or in the robotics and computer vision field, a concrete view of the synergy between these two areas. A second objective is to spur further fruitful exploitation of this connection.

The book is written at a level appropriate for an advanced undergraduate (or robot enthusiast) or graduate student. To make the text useful to a wide audience in these fields, the key concepts in robotics, computer vision, and cluster computing are introduced before being used.

I chose the algorithms and applications covered in the text because they were either easily accessible to a robotics person looking at cluster computing for the first time, or a cluster computing person looking at robotics for the first time. The first sections are necessarily therefore fairly simple robotics and cluster topics. However, the chapters build on each other and more complex robotics and cluster computing topics are covered in later chapters.

The reader will notice that each chapter ends with a bibliography section containing the references cited in that chapter. The reference numbers in the text refer to this section. I hope this system will offer the reader a convenience in locating references, though it is at the cost of some duplication of references. A merged bibliography is presented at the end of the text for completeness. The Appendices contain a summary of MPI related material from the text and from the MPI V2.2 specification.

It would be very difficult to provide a comprehensive list of robotics and computer vision algorithms and their cluster algorithm designs. My hope is that if a reader does not find the algorithm or analysis they need here, at least they find an entry point into the topic from which they can prototype their own solutions.

I have many to thank for their inspiration, insight and help in producing this text. I start by thanking Michael Arbib and Ken Overton for taking an engineering student with a distributed computing background into the Laboratory for Perceptual Robotics at UMass, kick-starting my interest in concurrency and robotics. I have been lucky to know and work with many brilliant and insightful professionals including Ron Arkin, Paul Benjamin, Tomas Brodsky, Eric Cohen-Solal, Teun Hendricks, MiSuen Lee, Yun-ting Lin, Tom Murphy, and many others. I owe a special debt of gratitude to Frank Hsu, who has been a mentor and friend in my years at Fordham University. Thanks are due to Stephen Fox, Gary Weiss and Arthur G. Werschulz for many comments on an early draft of the text. I also want to thank the many students who took my Parallel Computation course and my Robotics course at Fordham over the years and on which much of this material was dry-run including Sirhan Chaudhry, Sothearith Chanty, Jeremy Drysdale, Jose DeLeon, Giselle Isner, Andy Palumbo, Kiran Pamnany, and many others.

Finally, my family deserves all my thanks and more for graciously putting up with my many absences over the past few years while holed up in my office or lab assembling this text.

Damian M. Lyons
December 2010

Contents

List of Tables

List of Figures

Chapter 1

Introduction

There was a time when any book about robotics would need to start with a definition of what a robot is. This was, in some part, due to the fact that the robots of fiction had far outdistanced actual robots. Few people had seen an actual robot and authors needed to be mindful of the distractions, perhaps disappointments, which this could pose for their readers. It would be premature to say we are completely beyond this situation; however, robots are now much more common in the media and appear, and occasionally appear to behave, more like their fictional counterparts.

For example, the robot on the left in Figure 1-1 is the Robonaut 2, a humanoid robot designed to assist astronauts work and explore in space. The robot on the right is the HRP-4C humanoid developed by AIST in Japan. Both of these look and, to a limited extend, act more like the humanoid robots of fiction than like the industrial robots that have been with us since the late 1950s when George Devol and Joseph Engelberger founded the first robotics company, Unimation. Figure 1-2 shows a Puma® 560 robot arm, a robot built by Unimation for industrial applications such as welding, assembly, packing pallets and so forth.

1.1. Robots

Robotic technology has been limited for much of its history to these kinds of industrial applications. The surrounding environment in which the industrial robot operates, that is to say the robot's "world," can be tightly controlled. The objects that the robot needs to manipulate or interact with

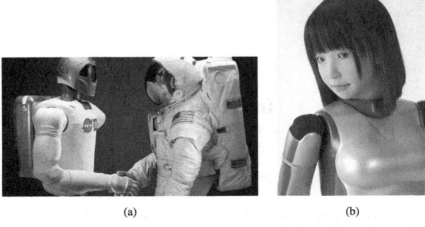

(a)   (b)

Figure 1-1:   (a) Robonaut 2 (R2) (reprinted with permission from NASA); (b) Humanoid robot HRP-4C (reprinted with permission from AIST, Japan).

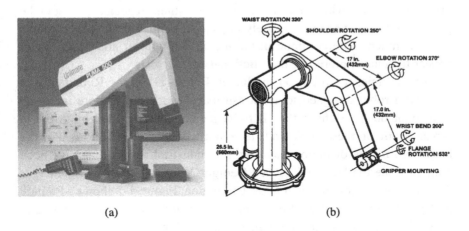

(a)   (b)

Figure 1-2:   Unimation Puma® 560 industrial robot.

can be precisely defined, and their timing and locations prearranged with a high degree of precision. In this way, the environment can be restricted to situations that the robot sensors, actuators and controller can handle quickly and successfully.

(a)   (b)

Figure 1-3: (a) iRobot 510 PackBot® (reprinted by permission of iRobot); (b) da Vinci® Surgical System (reprinted by permission of Intuitive Surgical Inc).

Robotic technology is now spreading quickly into other, new application areas, including:

- **Military robots** such as the Reaper, Predator and Gnat unmanned aerial vehicles (UAV's) from General Atomics, the Talon® unmanned ground vehicles (UGV) from Foster Miller and the PackbBot® from iRobot.
- **Professional service robots** such as RobotWatch's OFRO® security robots, Pipe and tunnel cleaning robots from RedZone Robotics, the da Vinci® surgical robot, and robot guides in hospitals and museums;
- **Robots for the home consumer market**, such as the Sony AIBO® and QRIO®, iRobot Roomba® and Scooba®; and,
- **Robots for the educational market** such as Lego Mindstorm®, Stiquetto™, the HandyBoard, etc.

In many of these new applications areas, the environment in which the robot operates cannot be as tightly controlled as it can be in manufacturing applications. Therefore the robot needs to be capable of sensing its environment and exercising some degree of *autonomous action* if it is to deal effectively with the unknowns it may face. A security robot, for example, cannot predict where, when and what objects or people it will come across. It needs to invest sensor and processing time into monitoring its environment in a way that manufacturing robots typically don't. It needs to select appropriate actions to ensure it achieves its task objectives without damaging itself or others.

The expansion of robotic technology into this wide panoply of new and demanding application areas is fueled by improved battery, motor, sensor and computation technologies, providing the necessary higher performance for a lower cost.

1.2. Cluster Computing

Cluster computing is an approach to achieving high performance, high reliability or high throughput computing by using a collection of inter-connected computer systems. The first Beowulf cluster was built by Donald Becker and Thomas Sterling at NASA's Center for Excellence in Space Data and Information Sciences in 1994. Their goal was to build Commodity Off-The-Shelf (COTS) based systems to satisfy specific computational requirements. They called their cluster of 16 off-the-shelf DX4 processors "Beowulf," and that name has now come to denote the entire class of COTS based cluster machines. The goal of cluster

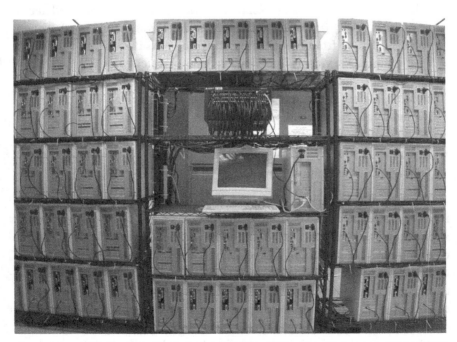

Figure 1-4: 128-Processor Beowulf cluster built by Donald Becker while he was at NASA/GSFC (reprinted with permission Michigan Technological University).

technology — supercomputing performance at off-the-shelf prices — is one that is directly in tune with the current needs of robotics and computer vision. Inexpensive, high performance computing is one crucial factor in building robots that can sense and respond effectively to events in this new range of unstructured environments. The additional computing power can be used to process sonar and visual information more quickly and/or more thoroughly; it can be used to recognize and track additional features of the robot's environment; and it can be used to pick better action strategies, even considering those based on past performance and observations.

The additional computing performance comes at the cost of developing robotic computing algorithms that can exploit the parallel computing resources available in a cluster. To understand why this is not trivial consider the physical task of digging a hole in your back garden. If it takes you a certain time to dig a $1\,m$ deep hole of $1\,m$ in circumference with a shovel, you can theorize that having a friend help you dig would cut the time in half. This speed-up will only be realized if you and your friend do not get in each other's way while digging! Just as a four-man rowing crew needs to coordinate its strokes to avoid oar collisions, two people digging a hole may need to coordinate digging strokes if the best speed-up is to be obtained. Similarly, in designing algorithms for cluster computation, significant effort may need to be put into the division of subtasks between cluster processors so as to get an effective speed-up.

Recently, multi-core systems have been adopted by processor manufacturers as a solution to improving processor performance. A multi-core Central Processing Unit (CPU) has several processor cores built into a single chip. Each core in a multi-core processor can function as a separate cluster processor. A Graphical Processing Unit (GPU) is a processor for handling graphical operations such as texture mapping, shading, image processing and so forth. A CPU can offload graphics tasks to one of these auxiliary processors. Multi-core technology has also been applied to constructing more powerful GPUs and these can also be incorporated into cluster computing.

In this book, we will look at how cluster technology can be leveraged to build better robots. More specifically, we will concern ourselves with the computer algorithms that control robot behavior. Algorithms and approaches in key areas of robotics and computer vision will be introduced.

These will broadly include areas such as map building, localization, target tracking, action selection and learning. Each will be introduced and cluster implementations for algorithms in these areas presented. It would be difficult to provide a completely comprehensive collection of robot and computer vision algorithms. The ones chosen for this text were selected for their accessibility to new robotics or cluster computing practitioners and as broad examples of classes of algorithms or approaches.

1.3.   Overview of the Book

Chapter 2 introduces the basic concepts of parallel processing and cluster computing. The focus is on understanding why cluster computing has become one of the dominant forms of parallel computing and how this relates to robotics. In addition to some more practical information about constructing clusters, this section lays the ground work for the mathematical models of computing and communication introduced in the next chapter.

Chapter 3 introduces the basic concepts in parallel programming, focusing specifically on message passing using the MPI standard. In addition to more practical information about compiling and running programs with MPI, this chapter also introduces several mathematical models used in later chapters to analyze the performance of parallel algorithms. In particular logarithmic time models are defined for the collective communication operations in MPI.

With Chapters 2 and 3 providing a cluster computing foundation, Chapter 4 begins the study of robot algorithms. It introduces the basic terminology for the motion of a wheeled mobile robot. The design process used in the remainder of the book is introduced here for the problem of *dead-reckoning*, that is, calculating the location of the robot based only on the motion commands transmitted to the robot. The design process consists of understanding how the data and operations can be *partitioned* on the cluster, carrying out the *program design* based on this partition, and *analyzing* the performance of the result. The MPI collective communication operations for *scattering*, *gathering* and *reducing* data are introduced here.

Chapter 5 looks at processing point data from sonar and laser sensors. In particular the Hough Transform algorithm is used to identify prevalent straight lines in a collection of points, a case that clearly shows the importance of balancing computation load on a cluster.

The important mobile robotics activities of localization (determining where the robot is with respect to a map) and mapping (building a map of the robot's surroundings from sensory data) form the basis of the material in Chapter 6. Map representation on a cluster is discussed and a parallel implementation of Monte Carlo localization is developed.

Computer vision algorithms can be one of the most computationally expensive algorithms that a robot needs to carry out. In Chapter 7, cluster implementations for iconic, or pixel-level computer vision operations are presented, as well as implementations for multi-scale versions of these operations. The important field of visual tracking is introduced and a parallel implementation of the Condensation algorithm using spatial histograms is presented.

Landmark selection and recognition can play a key role in allowing a robot to build a large-scale map. Chapter 8 continues the computer vision theme looking at issues involved in learning visual landmarks. Two unsupervised learning algorithms, K-Means and Expectation-Maximization, are presented.

Behavior-based approaches have had a profound impact on the field of robotics. Chapter 9 introduces the concept of behavior-based robotics and robot architectures. An MPI implementation of a typical, static behavior-based architecture is developed. The advantages of adding dynamic process creation and destruction to behavior-based robot architectures are then discussed. The new process control features of MPI 2.2 are used to build a dynamic behavior-based architecture.

Chapter 2

Clusters and Robots

There are many algorithms in robotics that have heavy computational demands. A typical example is that of analyzing a visual scene. This is computationally expensive because of the amount of data that needs to be processed: A single 24-bit color 640×480 camera image contains 900 Kbytes of data. A video sequence of 10 seconds of 10 frames per second (fps) of such images will require 88 Mbytes of data. Laser rangefinders also produce prodigious amounts of data. If the rangefinders are being used to help navigate a robot car such as *Junior*, Stanford University's entry (Figure 2-1) in the DARPA Urban Challenge competition,[a] then the data needs to be processed in a timely fashion.

A natural question to ask is whether adding additional computational resources would allow these algorithms to run faster and for larger problem sizes. In particular we can ask whether dividing the algorithm among several processors that can operate in parallel will allow an algorithm to run faster and for larger problem sizes. This question is not specific to robotics of course, and is addressed by the field of parallel processing (e.g., [3, 16, 19]).

2.1. Parallel Computation

There are several characteristics of robot platforms and robot programming that are key to understanding why parallel computing has a significant role in robotics [4, 7, 11]. Henrich and Honger [7] present a detailed list of such

---

[a] http://www.darpa.mil/grandchallenge/index.asp

Figure 2-1: Junior, the Stanford entry in the DARPA Urban Challenge (from [12]).

characteristics, a modified version of which is presented below as a series of perspectives or 'levels' (not always in the hierarchical sense).

(1) *At the level of control of actuators and sensors*: Each actuator and sensor is an independent physical device that can move (or sense) at the same time as other actuators or sensors.

(2) *At the level of groups of actuators or sensors*: Some actuators or sensors may be part of group always coordinated in their motions, for example, the fingers of a dexterous hand.

(3) *At the level of kinematic chains*: The 'arm' is a common electrome-chanical design template in robotics — an end-effector mounted at the end of a sequence of links. Each such chain can move at the same time as other chains in the mechanism.

(4) *At the level of functional modules*: Navigation, mapping, learning, task planning and other functional modules are somewhat independent of each other and their execution can be concurrent.

(5) *At the level of robot architecture*: Behavior-based robot architectures [1, 10] consist of networks of communicating behavioral modules. These networks can be considered as concurrent process networks.

(6) *At the level of multiple robot systems*: Each system can perceive, plan and act concurrently with other systems.

(7) *At the level of algorithms*: Each of the levels mentioned above will involve computation, and the algorithms employed in this computation may yield additional sources of parallelism [16, 19].

In this book we will focus on parallelism that is not necessarily linked to the physical design of the robot as are levels 1 through 3 above. Our primary focus will in fact be the last level, but we will cover issues at all four levels from 4 through 7.

2.1.1. *Parallel Architectures*

A sequential computer applies one operation to one unit of data at a time. Flynn's Taxonomy [3] is a traditional categorization of the possible parallel extensions to the sequential computer:

(a) SISD: Single Instruction, Single Data — the sequential computer.
(b) SIMD: Single Instruction, Multiple Data — a *vector* or array *processor* that carries out a single operation on multiple units of data at one time, or a *pipelined* array processor that carries out multiple operations on a conveyer-belt like data stream.
(c) MISD: Multiple Instruction, Single Data — the same data item is processes with multiple operations simultaneously. In a *systolic* array, data items are 'pumped' through multiple functional modules.
(d) MIMD: Multiple Instruction, Multiple Data: *Multiprocessor* or *multicomputer* systems operate on multiple data items with multiple operations simultaneously.

With the exception of pipelined processing, which is extensively present in commercial 'sequential' computers, the SIMD and MIMD categories have been relegated to niche computing applications such as digital signal processing, and the MIMD category has come to the fore. One reason for this has been the spread of personal computer technology. For example, a processor array typically consists of multiple arithmetic-logic units (ALUs) controlled by a single control unit. The Connection Machine (CM-1) [8] had as many as 65,536 simple processors sequenced by a control unit. However, this approach really only makes sense if the cost of the control unit is high. Thanks to PC technology, control units are relatively inexpensive. A second reason for MIMD dominance is the difficulty that SIMD and MISD designs have in riding the technology wave [15] — that is, in incorporating the very latest technology. New processor designs appear regularly and are quickly available in commercial computer platforms. The MIMD class consists of multiprocessors and multicomputers;

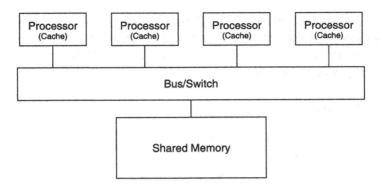

Figure 2-2:   Generic symmetric multiprocessor architecture.

most if not all of the literature on clustering concerns these architectures [2, 9, 15–17].

2.1.2.   *Multiprocessor*

A multiprocessor system consists of multiple processors that share a common memory system. The same memory address on any processor refers to the same physical memory location. Quinn [16] distinguishes *centralized multiprocessors* — which share a common memory — from *distributed multiprocessors* where memory is distributed on each processor and communication is via a network. The centralized multiprocessor is also known as a *symmetric multiprocessor* (SMP). Figure 2-2 shows a generalized architecture for an SMP. Each processor has equal (symmetric) access via a bus or switch to the single common memory — these designs are also called *uniform memory access* (UMA) architectures.

Cache coherence is a key issue in multiprocessor systems, and some form of snooping bus protocol is usually used to ensure that processors all see the same memory contents. Multicore processors leverage the SMP architecture. The Intel Xeon® E5462[b] is a quad-core processor intended for high-performance server and workstation markets. The multiple cores can be considered as multiple processors, but since they share the same physical die, interprocessor communication is easier. For example, the cores on the

---

[b]http://ark.intel.com/Product.aspx?id=33084

E5462 have separate L1 cache but share L2 cache. The E5462 also supports a dual independent bus architecture, allowing two quad-core processors in a system. A dual-E5462 is essentially an eight-processor SMP.

2.1.3. *Multicomputer*

A multicomputer system is a collection of separate processors with their own memory and I/O subsystems connected by a communication network. Processors do not share a common memory framework. The same address on different processors refers to different physical memory — a disjoint memory space. A crucial aspect of a multicomputer is the communication network, the mechanism that allows the individual computers to interact with each other.

The connection network can typically be either a *shared* network or *switched* network:

- In a shared medium network, a processor transmitting data can be heard by all other processors. Before transmitting a processor will need to 'grab the network' typically by waiting until the network is free, and then transmitting while listening for interference — a collision — from another transmitting processor. The 1970's radio-based ALOHAnet was a shared network. Ethernet is also a shared medium network and it inherits its core mechanism for grabbing the network from ALOHAnet: Carrier Sense Multiple Access with Collision Detection (CSMA/CD). It's easy to broadcast and multicast in a shared medium network.
- The switched medium network supports making a connection between any two processors. In a switched network, two processors who need to communicate can do so in a 'point-to-point' fashion without any other processors hearing. An advantage therefore of a switched medium network is that many pairs of processors can communicate concurrently.

One design option for a switched network is a cross-bar switch. However, the complexity of a cross-bar switch increases with the square of the number of processors. There are other network designs or topologies such as binary tree networks, butterfly networks, hypertree networks, hypercube networks and so forth [16]. Selecting a network topology for a specific application involves looking at the characteristics of each of these designs. They can be categorized in terms a number of communication metrics such

7. Application	Network process to application
6. Presentation	Data representation, encryption and decryption
5. Session	Interhost communication
4. Transport	End-to-end connections and reliability, flow control
3. Network	Path determination and logical addressing
2. Data Link	Physical addressing
1. Physical	Media, signal and binary transmission

Figure 2-3:   The seven-level OSI communication model.

as the largest distance — number of adjacent connections or edges — between any two processors, how many edges per processor, how easy it is to split the network into two disjoint parts, and others.

The Open Systems Interconnect (OSI) model views a network as a hierarchical sequence of seven levels (see Figure 2-3). The lowest levels concern the protocol of the physical medium such as the electrical and mechanical conventions. Higher levels specify addressing, flow control, reliability, encryption and so forth. Shared medium networks such as Ethernet suffer from a collision domain problem: the more processors that are connected to a shared medium, the more potential collisions that can occur, which will impact communication times.

A *Hub*, or repeater, is a device that makes a physical interconnection between networks. It operates at the physical level of the OSI model. Adding more processors to a shared network using a hub means that there will be more collisions, and it will become more difficult for processors to communicate. A *Router* is a device that can connect two network segments while isolating the collisions on each segment. Ethernet routers, which started to appear in the 1980s, also allow additional message filtering and protocol interfaces between different network types. So a router operates at OSI level 3 or higher. A *Switch* is a device that makes a low latency, high bandwidth direct connection between any two communicating processors, without recourse to collisions. *Latency* is the 'start up' time associated with each communication and *bandwidth* measures the rate at which data can be transferred during communication. Ethernet switches, developed in the early 1990s, operate at OSI level 2, the data link level. Multiple processor pairs can communicate in parallel without interference.

For example, the Hewlett-Packard ProCurve 2848 switch has the following latency and bandwidth specifications [9]:

Table 2-1: Hewlett-Packard ProCurve 2848 switch specifications.

ProCurve 2848	Latency	Bandwidth	Num. Ports	Backplane
100base-TX	11.8 μs	100 Mbps	48	96 Gbps
1000base-Tx	4.4 μs	1000 Mbps	48	96 Gbps

An important characteristic to note in switches is *oversubscription*: a switch is oversubscribed if the backplane bandwidth is less than the bidirectional bandwidth of all the ports. In this case, the bidirectional bandwidth of all ports is $2 \times 1000 \times 48$ Mbps or 96 Gbps, so the switch is not oversubscribed. The HP ProCurve 2924 has the following specifications:

Table 2-2: HP ProCurve 2924 switch specifications.

ProCurve 2924	Latency	Bandwidth	Num. Ports	Backplane
1000base-TX	3.7 μs	1000 Mbps	20	115 Gbps
10Gbase-CX4	2.1 μs	10 Gbps	4	115 Gbps

The bidirectional bandwidth is $2 \times 1000 \times 20 + 2 \times 4 \times 10000$ Mpbs or 120 Gpbs, so the switch is slightly oversubscribed in this configuration. There is much, much more that can be said about multiprocessors and multicomputers; however, this book is about using clusters, which are a specific kind of multicomputer, for robotics and computer vision algorithms.

2.2. Clusters

Definitions of a computational cluster abound. The following, from Lucke [9], is very general:

A closely coupled, scalable collection of interconnected computer systems, sharing common hardware and software infrastructure, providing a parallel set of resources to services or applications for improved performance, throughput or availability.

Clusters have a number of advantages over not only the SIMD and MISD architectures from Flynn's Taxonomy, but also over multiprocessor architectures [9]:

(1) **Scalability:** Cache coherence becomes expensive in larger SMPs and imposes a limit on the number of processors. The next step is to break the design into a cluster of SMPs.
(2) **Memory:** There are limits to the amount of RAM that a single SMP can support.
(3) **Cost:** A cluster of smaller SMP systems may end up being cheaper to build than a larger monolithic SMP because of the availability of commodity components.

2.2.1. *Terminology*

A commodity cluster is a cluster built from off-the-shelf components rather than specially designed processors and interconnects. The first *Beowulf cluster* was built by Becker and Sterling at the Center for Excellence in Space Data and Information Sciences in 1994 [5]. Their goal was to build Commodity Off-The-Shelf (COTS) based systems to satisfy specific computational requirements.[c]

A computational cluster in this sense is not the same as a *Cluster of Workstations* (*COW*) or *Network of Workstations* (*NOW*). The COW and NOW configurations are widely used for distributed problem solving applications. One such example is SETI@home, which uses a distributed network of 3 million workstations to do signal analysis of radio telescope data looking for evidence of intelligent life elsewhere in the cosmos. This is not a cluster application since the 3 million workstations belong to people who have first call on their use. SETI@home exploits idle processor time on these machines. *Grid computing* is another distributed processing paradigm. Grid computing differs from cluster computing in that it typically concerns a set of heterogeneous systems communicating via the internet and typically with a functionally distributed design.

---

[c]http://www.beowulf.org/overview/history.html

2.2.2. *Cluster Configuration*

There are three common configurations for a cluster:

(1) *High Availability (HA)*: The cluster resources are used to ensure that services are provided with high reliability by redundancy. A node failure is detected by another node in the cluster, which then assumes the failed node's IP and takes over the service.

(2) *High Throughput (HT)*: The cluster resources are used to complete the largest number of jobs in the smallest time. Typically the jobs are independent. "Compute Farms" are an example high throughput cluster application.

(3) *High Performance (HP)*: The cluster resources are used to complete a single task in the fastest time. This configuration is commonly used in scientific computing.

We will focus specifically on high performance applications for robotics and computer vision.

The simplest COTS cluster can be constructed by connecting a set of similar computers together. For example, the HP ProLiant DL160 server has dual, quad-core Intel Xeon® processors and can be treated as an SMP cluster on a board. When connecting multiple computing nodes together, it's important to use a *switch* rather than a router or hub, since this will provide low-latency parallel point to point communication. It may be advantageous to use UDP rather than TCP, if the cluster network is reliable.

In a *symmetrical cluster*, every node has the same operating system and the same resources. It doesn't make any difference which node a computation runs on. This means that parallel algorithms can be developed and implemented considering only the problem constraints and unimpeded by concerns over where the computation should occur on the cluster. In an *asymmetrical cluster*, nodes differ in their capabilities. Early clusters were distinctly asymmetrical, with a 'full-service' front-end machine to handle user interactions and a set of simpler back-end machines to handle the computation. Some asymmetry is probably required in the design of any cluster: a 'head' node might be connected both to the cluster switch via one of its network interface cards, and to an outside LAN for user interactions via another; and, a file server node might provide an NFS file service to the cluster.

Allowing nodes to easily communicate with each other will typically require file sharing and the ability to run remote processes. In Linux, this can be as easy as NFS (Network File System) and RSH/RLOGIN; however, RSH is not a very secure option, SSH is a better choice. Once these services are available, nodes can communicate using TCP/IP sockets. There are also a number of tools available that simplify configuring and using clusters. Our focus in this book will be on the algorithmic aspects of clusters in robotics and computer vision, so the tool list below is somewhat abbreviated. More 'how-to' details on these tools can be found in [2, 5, 9, 15].

2.2.3. *Programming the Cluster*

Parallel Virtual Machine (PVM)[d] is a message-passing library developed in the late 1980s at Oak Ridge National Labs for use with C/C++ and Fortran. It provides an API that allows programmers to spawn parallel processes and to communicate by message passing between these processes.

The Message Passing Interface (MPI)[e] standard was developed by a committee of parallel computing vendors and users. It started with MPI 1.0 in 1994 and is still undergoing active development (MPI 2.2 was released Sept 2009). MPI originally focused on the single program multiple data (SPMD) model with a static set of processes. However, dynamic process control was added to MPI in later versions.

OpenMP[f] differs from both PVM and MPI in focusing on shared memory systems. It employs a Unix *Fork-Join* model to create teams of parallel threads to carry out parallel computation. The threads share the same memory space (they have private space as well) and use synchronization primitives to implement mutual exclusion and communication.

Message passing is the more natural paradigm for cluster computing. MPI probably has the edge in terms of standardization and usage among message-passing libraries and will be covered in more detail in Chapter 3.

---

[d]http://www.csm.ornl.gov/pvm/
[e]http://www.mcs.anl.gov/research/projects/mpi/
[f]http://openmp.org/wp/

2.2.4. *Configuring the Cluster*

OSCAR[g] (Open Source Cluster Application Resources) is a collection of software for configuring and using commodity clusters. OSCAR will set up the head node of the cluster and will build the image to install on the other cluster nodes.

Rocks[h] is a more full service, cluster configuration tool that uses the Red-Hat Linux distribution appropriate for medium to large clusters.

LinuxPMI[i] *is* an extension of the Linux kernel that allows processes to migrate transparently among different machines within a cluster so that workload can be balanced. This continues the *openMosix* project which halted at version 2.6 in 2008. Installing LinuxPMI requires being comfortable with making patches to the Linux kernel.

2.2.5. *Simple Cluster Configuration with OpenMPI*

It is possible to set up a cluster without using any of these tools. Let us assume we are starting with a collection of identical computers running Linux and connected via an Ethernet switch. We will assume that the switch is set up to use the private IP address space *10.10.1.x* where x is the machine number in the cluster. *OpenMPI* is an open-source implementation of the MPI-2 standard. It can be downloaded along with documentation from *http://www.open-mpi.org/*. Each computer that is to be part of the cluster will need to have OpenMPI installed.

Once OpenMPI is installed on a computer, MPI programs can be compiled and can be run on that machine. If the machine is a multiprocessor and/or a multicore machine, then MPI can be used to build parallel programs. On a single processor system, MPI can be used to emulate parallel processing.

However, to run MPI programs on the entire cluster, it is necessary to set up the collection of machines so that one machine, the *head (master) node (machine)* for the cluster, is capable of running programs on all other *(slave) nodes (machines)*. Since the machines are identical, it will be possible to run executables generated on the head node on every other

---

[g]http://www.csm.ornl.gov/oscar/
[h]http://www.rocksclusters.org/
[i]http://linuxpmi.org/trac/

node. We will consider setting this up for one user *robotmpi*; the remaining steps will need to be done for every user of the cluster.

Each cluster machine will need to have NFS (Network File System) installed and configured. The *robotmpi* user will have a home directory on the head node and that directory is exported to all the other machines. For example, if the home folder is */home/robotmpi* then the following line should be added to the head node's */etc/exports* file:

/home/robotmpi   10.10.1.0/255.255.255.0 (rw, sync)

The command *exportfs –a* executed on the head node will export this shared folder to all other machines on this *10.10.1.x* network. The other machines will need to include a line in their */etc/fstab* to mount this directory as their local */home/robotmpi*. Note that for more efficiency, a separate switch can be added to take the NFS traffic off the cluster switch. Each machine would then need two network interface cards, one for NFS traffic and one for inter-node communications.

Next we need to allow the *robotmpi* user to login to each of the cluster machines. For this we will use *OpenSSH*, a program that provides secure remote access and which can be downloaded from *http://www.openssh. com/*. Every node needs to have the OpenSSH server installed and running, and the head node will need to have the OpenSSH client installed.

The *ssh-keygen* and *ssh-add* commands, parts of OpenSSH, support the generation of public/private key pairs which allow *robotmpi* to login from the head node to any other node without a password. This is the last step needed. MPI program can now be run from the head node on every node in the cluster.

2.2.6. *Connecting the Cluster to the Robot*

Although cluster use in robotics is not prevalent yet, given the growth in multiprocessor, multicore systems, it is a reasonable prediction that it may soon be so. One approach to using a cluster for a robotics application is as an off-board (i.e., not on the robot) computational resource. This is shown in Figure 2-4.

The head node of the cluster communicates via WiFi with one or more robots in a team. The robots carry their own computational resources, but reserve these for carrying out actuator and sensor control and any

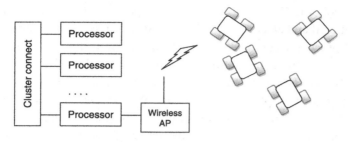

Figure 2-4: Off-board cluster configuration.

motor behaviors that require fast reaction time. Tasks that do not require a fast reaction time and that have a high computational load are processed on the cluster. Examples of tasks that should be carried out onboard in this example are motor control, low-level sensory processing, obstacle avoidance and local navigation (that is, for the immediate vicinity of the robot). Examples of tasks that could be carried out on the cluster include sensor fusion, mapping, landmark learning and global navigation (that is, for the entire area in which the robot is working, including other robots).

Advantages of this approach include the lighter computational load on individual robots and the ease of integrating multiple robot information. Disadvantages include the added communication time to ship information back and forth from each robot and the single point of failure issue of a centralized resource. Of course this latter point could be ameliorated by including some High Availability characteristics in the cluster.

Another approach is to equip each robot with a cluster. This is not as novel an idea as it may sound. Navlab 1 [6], a self-driving Chevy van built at Carnegie-Mellon University in 1986, carried 5 racks of computer equipment onboard (and moved at a top speed of 20 mph). The 1990's Navlab 2 also carried 5 onboard computers and had a top speed of 70 mph. Junior, the Stanford entry in the DARPA urban challenge carries two quad-core processors. Of course, onboard computers do not have to be configured as clusters. Functionality can be statically distributed across computers resources: one machine assigned to processing visual input, another to trajectory control and so forth. However, in this book, when we refer to an onboard cluster, it will be in the sense we have used up until now, of a set of parallel resources that can be used for a task.

The Beobot [13] project takes the concept of COTS design to both the computational and electromechanical design of a robot. The Beobot carries two dual-processor computers connected by gigabit Ethernet, a cluster that provides sufficient computational power for complex visual tasks. The Segway-Soccer Brain-Based Device (SS-BSD) [18], a modified Segway scooter designed to play mixed human/robot soccer, carries an onboard cluster of seven Pentium-IV compact PCs connected with gigabit Ethernet. The cluster is used primarily to run the simulated nervous system, based on a simplified model of a vertebrate nervous system, that controls the scooter.

There are two configuration approaches we can take to connecting an on-board cluster to robot hardware, shown in Figure 2-5. An asymmetrical onboard cluster (Figure 2-5(a)) is connected to the robot via a single processor, and looks similar in many respects to the off-board configuration (Figure 2-4). The head node on the cluster carries out all the interactions with the robot. A symmetrical cluster (Figure 2-5(b)) has the same connectivity to the robot on every processor of the cluster. This has all the advantages discussed previously for symmetrical clusters, but leaves open

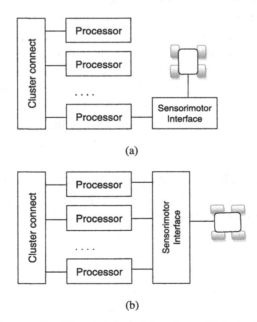

(a)

(b)

Figure 2-5: (a) Asymmetric on-board cluster configurations. (b) Symmetric on-board cluster configurations.

the issue of potential conflict between processors in controlling the robot. In fact, it is advantageous to take arbitration of robot resources out of the cluster and into the domain of robot architecture. We will discuss this in detail in Chapter 9.

2.3. Summary

This chapter briefly introduces the field of parallel architectures and parallel computing. Enough detail is presented to place the concept of the computational cluster in context, and as a basis for the mathematical models of processing and communication introduced in later chapters. Cluster computing is also introduced. There are a great many resources available on the web for building and using clusters, and pointers to this body of information are presented.

Cluster use in robotics is not prevalent yet. However, given the growth in multiprocessor, multicore systems, it's reasonable to expect that it soon may be. With this in mind, we examined how clusters can be connected to robot equipment. We divided this into on-board and off-board configurations. For onboard configurations, we extended the definition of symmetric and asymmetric clusters to cover connection to the robot. While symmetric onboard configuration have advantages in terms of simplifying the constraints on what processing is done where in the cluster, they leave open the issue of potential conflict between processors in controlling the robot. We point out this may be a good thing, since this arbitration is a key aspect of behavior-based robot architectures.

Several examples of Hewlett-Packard equipment were given in the chapter; this is not an endorsement of HP equipment — the equipment is used in the author's lab so it was an easy choice for examples.

References

1. Arkin, R.C., *Behavior-Based Robotics*, MIT Press 1998.
2. Bookman, C., *Linux Clustering*, New Riders 2003.
3. Flynn, M., Some computer organizations and their effectiveness, *IEEE Transactions on Computing*, Vol. C-21, pp. 948, 1972.
4. Graham, James H., Special computer architectures for robotics: Tutorial and survey, *IEEE Trans. on Robotics and Automation*, vol. 5, no. 5, 1989.
5. Gropp, W., Lusk, E., and Sterling, T., *Beowulf Cluster Computing with Linux*, MIT Press 2003.

6. Goto, Y. and Stenz, A., The CMU System for Mobile Robot Navigation, *IEEE Int. Conf. on Robotics and Automation*, pp. 99–105, 1987.
7. Henrich, D. and Honiger, T., Parallel processing approaches in robotics, *IEEE International Symposium on Industrial Electronics (ISIE'97)*, Guimarães, Portugal, July 7–11, 1997.
8. Hillis, W.D., *The Connection Machine*, MIT Press 1989.
9. Lucke, R.W., *Building Clustered Linux Systems*, Prentice Hall 2005.
10. Lyons, D.M., Representing and analyzing action plans as networks of concurrent processes, *IEEE Transactions on Robotics and Automation*, June 1993.
11. Lyons, D.M. and Arbib, M.A., A formal model of computation for sensory-based robotics, *IEEE Transactions on Robotics and Automation*, 5(3), June 1989.
12. Montemerlo, M. *et al.*, Junior: the Stanford entry in the urban challenge, *Journal of Field Robotics* Volume 25, Issue 9, (September 2008).
13. Mundhenk, N., *et al.*, Low cost, high performance robot design utilizing off-the-shelf parts and the Beowulf concept, The Beobot project. *Proceedings, SPIE Intelligent Robots and Computer Vision XXI: Algorithms, Techniques, and Active Vision*, (Providence RI, 28–29 October 2003).
14. Pierre Sermanet, Raia Hadsell, Marco Scoffier, Matt Grimes, Jan Ben, Ayse Erkan, Chris Crudele, Urs Muller and Yann LeCun, A Multi-Range Architecture for Collision-Free Off-Road Robot Navigation, *Journal of Field Robotics*, 26(1):58–87, January 2009.
15. Pfister, G., *In Search of Clusters*, 2nd Ed., Prentice Hall, 1998.
16. Quinn, M.J., *Parallel Programming in C with MPI and OpenMP*, McGraw-Hill 2004.
17. Sloan, J., *High Performance Linux Clusters*, O'Reilly 2005.
18. Szatmary, B., Fleischer, J., Hutson, D., Moore, D., Snook, J., Edelman, G. M., and Krichmar, J., A Segway-based human-robot soccer team, *IEEE International Conference on Robotics and Automation*, Orlando FL, 2006.
19. Wilkinson, B. and Alen, M., *Parallel Programming*. 2nd Ed., Prentice-Hall 2005.

Chapter 3

Cluster Programming

This chapter will introduce two key ideas. The first is the programming language we will use to build parallel algorithms for robot applications. The *Message Passing Interface* (MPI) will be our main tool for building parallel programs. MPI is an Application Programming Interface (or API for short) for expressing explicit parallelism principally in C/C++ and Fortran programs (through other MPI language bridges do exist). The second key idea is the mathematical models we will use to evaluate the time complexity of the programs we design. In particular we will introduce Quinn's models [13] for the MPI collective communication times.

3.1. Approaches to Parallel Programming

The design of an algorithm to solve a problem using the standard, sequential programming model involves developing a sequence of concrete steps that, when carried out, will lead to the correct solution for the problem [2]. Consider for example, the task of programming a robot equipped with a *dexterous hand* [14] to pick up a cup and drink from it.

A dexterous robotic hand is a robot mechanism, typically modeled on the human hand and capable of grasping a wide range of object shapes and sizes. Some dexterous robotic hands are also capable of executing fine motions with a grasped object. Two dexterous robotic hands based on the human hand model are shown in Figure 3-1.

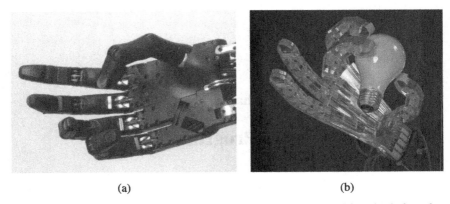

(a)   (b)

Figure 3-1:   Two dexterous robotic hands: (a) The C6M smart motorized hand from the shadow robot company (Copyright — The Shadow Robot Company — 2010); and, (b) The RAPHaEL (Robotic-Air Powered Hand with Elastic Ligaments) from the robotics and mechanisms lab at Virginia Institute of Technology (reprinted with permission, Dr. Dennis Hong, Director of RoMeLa, Virginia Tech).

The cup grasping task requires the following steps:

(1) Estimating the position of the cup.
(2) Setting the arm joint positions to bring the hand over to the cup.
(3) Estimating the size and orientation of the handle.
(4) Preshaping the hand as it approaches to the cup.
(5) Grasping the handle when the hand is sufficiently close.

Parallelism is attractive because it relaxes the sequence constraint, allowing more than one step at a time, and hence potentially reducing the overall time taken to solve the problem. In the reaching example above, the hand preshaping operations (steps 3 and 4) and the arm movement operations (steps 1 and 2) can be carried out simultaneously. However, parallel programming offers additional challenges beyond sequential programming, for example, the two parallel operations must coordinate sufficiently so that by step 5 the hand can grasp the handle (see Figure 3-2).

We will refer to the location at which, and mechanism by which, computational steps are carried out as a *processor*. The instructions executed on a processor will be called a *process*. The easiest case of algorithm design for parallel programming occurs when the steps to be taken on each processor are independent of the steps on the other processors. This case

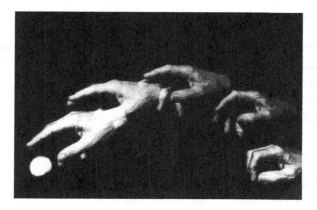

Figure 3-2: Overlaid time lapse photos of reaching and grasping a ball (Jeannerod *et al.* [8]).

is not much more difficult than sequential programming, but nonetheless it yields a tremendous *speedup*. Speedup is the ratio between parallel and sequential execution times; a more formal definition of speedup is introduced in the last section of this chapter. In the reaching example, the time taken in the sequential case is the sum of the times for steps 1 through 5. However, in a parallel implementation, the time taken is the maximum of the times for reaching (steps 1 and 2) or preshaping (steps 3 and 4) followed by the time for grasping. This is always at least as fast[a] as the sequential implementation and typically faster. When similar steps can be implemented on p processors, and there is no need for processors to interact, then the speed approaches a *linear speedup*, p.

The more common and more complex case occurs when the steps that need to be carried out on one processor must interact with the steps being carried out on another processor. Understanding this interaction is what makes parallel programming challenging because this interaction may also reduce the speedup and hence the benefit of parallel programming. In the worst case, the interaction is an additional overhead that causes the parallel program to run even slower than its serial counterpart.

Given the difficulty of writing parallel programs, it is not surprising that many programming languages have been designed to help the programmer build correct and efficient code [15][11][5]. In general we can divide these

---

[a]In the case when the hand is already 'preshaped' to the object and/or the hand is at the cup when motion begins.

languages into two groups:

(1) *Implicit Parallel Languages*: Languages that do not require the programmer to think explicitly about parallelism, but which automatically identify and implement any parallelism in the program.
(2) *Explicit Parallel Languages*: Languages that allow the programmer to specify what portions of a program need to be carried out in parallel and how the parallel sections of code should interact with each other.

Examples of the first class of languages include SISAL [5] a functional parallel programming language, and ZPL [3] a data-parallel, array programming language. While implicit parallel languages are clearly the choice to simplify algorithm specification, up to now these are mainly niche languages.

Examples of the second class of languages, the class in which parallelism is expressed explicitly by the programmer, can be further divided based on the model of parallelism they employ:

(1) *Shared-Memory Model*: This model assumes that all the processors share a common address space and can communicate with each other via shared memory data structures.
(2) *Message-Passing Model*: This model assumes that processors are connected to each other by a communications network and can communicate with each other by sending messages over the network.

Shared memory languages employ a mechanism such as a semaphore, critical region, mutex, or monitor to arbitrate between multiple processors accessing a block of common memory. OpenMP, a language developed for Symmetrical Multiprocessor Systems, is an example of the first class of language [4]. One of the earliest message passing languages was Hoare's CSP (Communicating Sequential Processes) [7], on which the transputer language OCCAM [9] was based.

Another important categorization for explicit parallel languages is whether the language is a full programming language or whether it is a library and an API (Application Programmer Interface) for an existing language. OpenMP is not a full programming language. It is a library and

API that can be linked with programs in an existing language to express parallel programs with shared-memory communications. Similarly, MPI (the Message Passing Interface) [11] is an API for expressing message-passing parallelism. Both of these popular APIs can be used to build parallel programs in C/C++ and Fortran as well as other languages to which the API is bridged (for example, Java and Python). PVM (Parallel Virtual Machine) is another message-passing API, released in the 1990's from Oak Ridge National Laboratory [5]. An advantage of the API approach to explicit parallelism is that it builds upon the existing collection of algorithms and community of users for a language.

In this book, we use MPI in a C language environment. The sections of code in the book are in C and we assume a basic level of familiarity with the C language. The C language API for MPI will be introduced and explained as we go along; no prior familiarity with that is necessary. A review of the application programming interface for MPI, as used in the book, is given in Appendix IV.

3.2. Programming with MPI

MPI is a standard created by the MPI forum.[b] The forum is a collection of parallel computer vendors, academic researchers and industry practitioners with expertise in parallel computation. The objective of the forum is to assemble the best ideas from various message-passing models into a single API that can be used in a variety of problems on a variety of parallel hardware.

The first MPI standard, referred to as MPI-1, was released in 1994. The second MPI standard, MPI-2, extended MPI-1 with feedback from users on MPI-1 and was released in 1997. Updates are released regularly. Implementations of the MPI standard are widespread. Indeed, if you are running a Linux operating system, it's possible that you already have MPI installed, since it is often bundled with Linux distributions. There are several implementations of MPI that you can download, including OpenMPI (available at www.open-mpi.org) as well as MPICH (available

---

[b]www.mpi-forum.org

at www.mcs.anl.gov/mpi/mpich1). The version of MPI used in this book is OpenMPI.

3.2.1. *Message-Passing*

Point-to-point communication between processes by message-passing involves the following steps:

(1) The *sending process* packages its data and addresses it to the receiver.
(2) The *sending process* requests to transmit its message to the receiver.
(3) The *receiving process* checks if there is a message waiting for it.
(4) The *receiving process* unpackages the message.

Several issues are raised immediately by this. First of all, each process needs to have a unique name. In MPI, the name of the process is an integer value called its *rank*. When a sending process $p1$ transmits a message to another process $p2$, it may want to wait until that message has been received before it proceeds to its next step. That is, it pauses at step 2 above until it determines that step 4 has occurred. This is called a *synchronous* send operation. In a similar fashion, the receiving process $p2$ may pause at step 3 if there is no message waiting, and not proceed to step 4 until a message has been transmitted to it. This is called a *synchronous* receive operation.

The *rendezvous* of $p1$ and $p2$ is the event that occurs when $p1$ executes synchronous send and $p2$ executes synchronous receive. During this event, data is passed between $p1$ and $p2$, but also, both processes are temporally synchronized with respect to each other. Going back to our example of reaching to grasp an object, note that the location of an object using, for example, a visual sensor, must be completed before reaching can begin. One way to model this information flow is as a synchronous send of the object location by the visual analysis process and a complementary synchronous receive by the reach process (Figure 3-3).

Of course, there may also be examples where no synchronization occurs, only the passing of data is required. In that case, neither $p1$ nor $p2$ need to pause; $p1$ sends the message and then proceeds immediately to its next step. $p2$ checks for a message and, if one is waiting, it reads it and proceeds to its next step; if one is not waiting, $p2$ just proceeds to its next step anyway. These are called *asynchronous send* and *asynchronous*

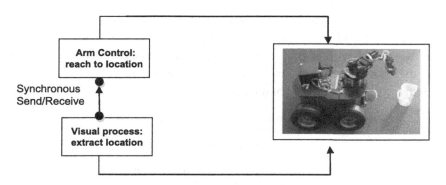

Figure 3-3: Synchronous send/receive for object grasping.

receive. In our reaching example, if the position of the target was continually changing (e.g., the target was a moving object) then it would make sense to transmit and receive that information asynchronously in a loop so that the latest position is always made available to the reach operation.

3.2.2. *Single Program Multiple Data (SPMD) Model*

Parallel programs are built with MPI using the Single Program Multiple Data model: The same program executes on every processor in the parallel machine. However, it is rare that having every processor execute exactly the same code will be what a programmer wants. In the MPI SPMD model, a process can tell on what processor it is executing by making a call to determine its rank. Recall that the rank acts as the ID for the process. A conditional statement can then be used to specialize the operations performed by that particular process.

The least convenient case for SPMD is when every process has completely different activities. In our reaching example, we would need code similar to:

```
myID = get_my_rank();
if (myID == 0)
        do_visual_module();
else if (myID ==1)
        do_reaching_module();
else .....
```

In many cases, parallel algorithms can be described in a "manager/ worker" [5] paradigm. The manager process partitions activities and data out to a set of worker processes; manager and worker processes operate together on the activities and data; finally, the manager process gathers together the results of the computation.

```
myID = get_my_rank();
if (myID == 0) /* the manager or 'root' process*/
    send_out_data_and_activities();

process_some_portion_of_data_and_activities();
/* manager & worker */

if (myID == 0)
    gather_results();
```

This paradigm is much easier to code in the SPMD model, and by and large, this is the paradigm we will leverage for building cluster implementations of robot algorithms whenever possible. In fact, the SPMD model is not a hard constraint in using later versions of MPI, MPMD (Multiple Program Multiple Data) programs can also be specified, as we will see.

In later chapters, when we develop parallel programs, we generally begin our development of a program by looking at how the data and activities can be spread among the processors in the cluster. We call this *partitioning*. Different choices of partitioning have different implications for the amount and kind of processing each processor will need to do as well as for the interactions between the processors. The last section of this chapter introduces some mathematical models that can be used to help make decisions about what kind of partitioning will yield the best performing algorithm.

3.2.3. *Collective Communication*

MPI provides a class of operation that is particularly useful in the "manager/worker" paradigm. A *collective communication operation* is an operation in which all processes participate. Compare this to the point-to-point send/receive shown in Figure 3-3, where only $p1$ and $p2$ participate in the communication operation. One common example of a collective communication operation is a *broadcast*. In a broadcast operation, one process (typically the 'root' or manager process) participates by providing the information to be sent, and all the other processes participate by receiving that information.

The following are some other common kinds of collective communication operations:

- **Scatter**: In a scatter operation, the root process distributes some portion of a data buffer to all other processes. For example, if there are p processes in total, including the root or manager process, and a data buffer with n items to be scattered, then, assuming n is a multiple of p for clarity, we can scatter n/p items to each processor as their balanced share of the workload. The first process gets the first n/p sized block of data. The second processor gets the second block, and so forth. (Later we will discuss the case where n is not a multiple of p.)

- **Gather**: A gather operation is similar to a scatter in reverse — the root process receives some portion of a data buffer from each other process. Continuing the previous example: after scattering the n/p data items, each process operates on those items, transforming them in place. The root then gathers the n/p block of data from each process and reassembles it in the correct location in its data buffer. The first n/p sized block of data in the buffer comes from the first process, the second block comes from the second process, and so forth.

- **Reduce**: A reduction operation takes data from each process and stores the result of applying a binary associative operator between all data items. There are various reduction operators available in MPI including summation, multiplication, maximum, and so forth. For example, let us say that a range of n/p values has been scattered to every process, and that each process has identified the maximum value in its n/p range. A reduction operation using the binary, associative operation MPI_MAX can be used to calculate the maximum of each process's maximum value and store that in the root process.

Collective communication operations can be implemented in an efficient way. One implementation approach for a collective communication

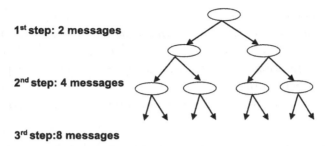

1st step: 2 messages

2nd step: 4 messages

3rd step:8 messages

Figure 3-4: Implementation of a collective communication operation using a logarithmic time method.

broadcast operation is for the root process to execute $p-1$ send commands, one to each of the $p-1$ other processes. This would take a time linearly proportional to the number of processes. However, a better approach is shown in Figure 3-4. The root process sends its message to two other processors; they in turn each send their message in parallel to two more, which in turn, in parallel, send their messages to two more. The time for all these 'waves' of messages to complete is proportional to $\log p$ rather than p (as is shown in Figure 3-4). All the collective communication operations can be implemented in this logarithmic style.

3.3.   Compiling and Running MPI Programs

Specific MPI commands will be introduced in subsequent chapters as we look at the solutions for a collection of robot programming problems leveraging the computational power of a cluster. We will start by briefly introducing here:

(1) The 'template' for a typical C program using MPI.
(2) The steps to compile and execute C/MPI programs from the Linux shell.

The code section below shows a template for a C program that uses MPI. Note that all the MPI commands begin with the characters "MPI_" and that some of them have the argument "MPI_COMM_WORLD." This argument is called the *communicator* argument and it describes the collection of processes that are to participate in collective communication operations. The default communicator includes all processes.

```
#include <stdio.h>
#include <mpi.h>

int main (int argc, char **argv)
{   int id, p;

    MPI_Init(&argc,&argv);
    MPI_Comm_rank( MPI_COMM_WORLD, &id );/* rank of this proc*/
    MPI_Comm_size( MPI_COMM_WORLD, &p );/* how many proc*/

    /* the SPMD model program is here */

    MPI_Finalize();
    return 0;

}
```

The variable id is set to the rank of the process, and the variable p is set to the number of processes in the communicator — in this case the total number of processes.

Let us assume the above program is saved in a file *myfile.c*. The program can be compiled with the command:

mpicc myfile.c −o myfile

If the compilation is successful, the program will be available to execute in *myfile*. However to distribute the program to the processors in the cluster, a special execute command is required:

mpirun −np *p* myfile

This command makes one copy of the program for each processor from *1* through *p* and starts their execution. To know where each processor is, MPI needs access to a hostfile. This can be specified as an argument with mpirun −hostfile filename −np *p* myfile. The *hostfile* contains one line for each host in the cluster as well as the number of processes that the host can support. For example, a dual processor host can support two parallel processes; a dual quad-core processor can support eight parallel processes. In fact, MPI is happy to run more processes than processors — but this will not produce real parallelism of course. More details of the arguments and behavior of mpirun are given for reference in Appendix I.

3.4. Analyzing Parallel Computation Time

The analysis of algorithms is a well developed field of computer science [2]. Consider the problem of searching through a list of n values for the maximum. This operation requires testing each value one by one. In the *worst case* you will have to search to the end of the list to find the maximum value. The worst-case execution time will be linearly proportional to the number of values in the list. Of course, executing this on a machine with a fast (small) instruction time will result in a fast run-time, but on a slower machine, a slower run-time. Nonetheless, as the size of the list grows, the execution time will just increase linearly with the size of the list. We say that any algorithm that has a run time directly proportional to the size of the input has the same linear *computational complexity*. Any algorithm whose worst-case complexity is bounded above by n is said to be in the computational complexity class $O(n)$ — for short, it's said to be $O(n)$ or equal to $O(n)$. We say that $f(x)$ is bounded from above by $g(x)$ if positive constants c and $x0$ can be found such that:

$$0 \leq f(x) \leq cg(x), \quad \text{for all } x > x0 \text{ for small } x0.$$

Consider the problem of multiple robot landmark identification. Let's say a robot exploring an area stores a collection of n landmarks to help it map out the area. A second robot navigating through this area a little later, also stores a collection of n landmarks. The problem is to determine which of the first robot's landmarks correspond to which of the second robot's landmarks. A direct way to accomplish this is to compare each of the first robot's landmarks to all the second robot's landmarks. This will require a run-time of $n \times n$ landmark comparisons and is hence in $O(n^2)$. Polynomial complexity classes such as $O(n^2)$ are worse than linear complexity in that the computation time increases much more drastically for increases in the size of the input.

Returning to the searching example: If we can rely on the fact that the list of n items we are searching is sorted already by value, then the well-known *binary-search algorithm* will find our value in $log\,n$ steps and is in $O(log\,n)$. The computational complexity class $O(log\,n)$ is better than

$O(n)$ in the sense that the computation time increases less drastically for increases in the size of the input.

Note however that binary search is also in $O(n)$ since $log\, n$ is bounded above by n. It is also in $O(n^2)$ since $log\, n$ is bounded from above by n^2. The notation $\Theta(log\, n)$ denotes the computational complexity class that is bounded above *and* below by $log\, n$ — a tight bound on the complexity. Binary search is in $\Theta(log\, n)$ but not in $\Theta(n)$.

We have already seen an example of computational complexity applied to parallel computation. The collective communication operations implemented naively as an iterative loop of $p - 1$ sends have a worst-case complexity in $\Theta(p)$. However, leveraging parallel communication resources allows us to formulate a parallel algorithm for collective communications with a complexity in $\Theta(log\, p)$.

3.4.1. *Speedup and Amdhal's Law*

The simplest measure of parallel performance is the parallel speedup ψ — the ratio of sequential to parallel execution times:

$$\psi = \frac{T_{serial}}{T_{parallel}}$$

Amdahl's Law allows us to estimate the maximum speedup. Consider a program such as the reaching program in Figure 3-1. Some portion of this program can leverage parallel resources (the reach and the preshape can be done in parallel). The remaining portion is intrinsically serial (the final grasp can only occur after the reach and preshape have completed). Let f denote the fraction that is intrinsically serial. Therefore $1 - f$ is the fraction that can leverage the p parallel resources. We can say that

$$\psi_{max} = \frac{1}{f + (1 - f)/p}$$

Note that as p gets larger, the $(1 - f)/p$ term gets smaller, and, in the limit, as the number of processors goes to infinity, $\psi_{max} = 1/f$. For example, if we identify the serial component as 0.25, then the best speedup we can get is $\psi_{max} = 4$. In fact, since Amdahl's law does not take communication overhead into account, the best speedup may be a lot less than this value.

3.4.2. Communication and Calculation

The parallel execution time is made up of two pieces: the time taken to calculate on a processor and the time taken to communicate with other processors.

$$T_{parallel} = T_{cal} + T_{com}$$

Consider the problem we introduced when discussing collective communications: Finding the maximum of a large array of values. The serial time is $\Theta(n)$ of course. For the parallel implementation, recall first that the n values have been scattered, with every process receiving n/p values. Each process identifies the maximum value of the function in its n/p range. A reduction operation using the maximum operation is used to calculate the maximum of each of these values and store that in the root process. The calculation time on each processor is $\Theta(n/p)$. The scatter and reduction operations are each $\Theta(\log p)$.

Because T_{cal} tends to increase more quickly with problem size n than T_{com}, the speedup tends to be an increasing function of problem size. This is called *Amdahl's Effect*. Conversely for fixed size n the speedup tends to be a decreasing function of the number of processors.

Amdahl's Law assumes that the parallel fraction benefits from the p parallel processors without any overhead of communication. A more realistic measure is the Karp-Flatt metric [13], the so-called experimentally determined serial fraction e:

$$e = \frac{\frac{1}{\psi} - \frac{1}{p}}{1 - \frac{1}{p}}$$

The Karp-Flatt metric can be leveraged to understand whether the decrease of speedup with increasing p is due to a poor algorithm with limited scope for parallelism or due to parallel overhead.

3.4.3. Communication Models

We model the communication commands in MPI with a standard bandwidth and latency Hockney model [12]. When a message is sent — either

point-to-point or by collective communication — there is some *latency* or
initial time delay, associated with the transfer. Let us denote this by λ. We
assume that the communication channel has a *bandwidth* associated with
it — the most information that can be transmitted in a unit time. We denote
this by β. Thus if we wish to send a message of length n then it will take a
time

$$T_{com} = \lambda + \frac{n}{\beta}$$

The HP ProCurve 2900 Ethernet switch lists a latency of 3.7 μs at a
bandwidth of 1 Gbps, and 2.1 μs at a bandwidth of 10 Gbps. The Quadrics
QsNet(II) Elan switch, a specialized interconnection for High Performance
Computing (HPC) offers 24 ns latency at 1.3 Gbps. The Netgear GS608,
an inexpensive switch, has a latency of 1.2 ms at 10 Mbps and 15 μs
at 1 Gbps.

We adopt Quinn's approach [13] to model collective communication
operations. Consider scattering n items to p processors. In the first step,
the root process sends half of the list to each of two processes. In the
second step, each of these sends a quarter of the list to four processes,
and so forth. We can write the total communication time for this binary
tree implementation of scatter as:

$$T_{gather} = T_{scatter} = \sum_{i=1}^{\log p} \left(\lambda + \frac{n}{2^i \beta} \right) = \lambda \log p + \frac{n(p-1)}{\beta p}$$

Gather and scatter communication operations are the inverse (in time)
of one another, so this formula applies to both. Figure 3-5 shows a graph
of the estimated $T_{scatter}$ using this expression versus the actual $T_{scatter}$
measured as the average of 10,000 scatter operations for an array of 1200
floats on an HPC cluster.

The 16 processors used in this measurement were on two, dual processor, quad-core blades. For the graph labeled 'Measured S8' in Figure 3-5,
the first 8 processors were all allocated on the same blade, and the final 8
were on a second blade. (See Appendix I for specifying processor binding
options.) The first 8 blade cores have significantly better than predicted
performance. The latency and bandwidth of core-to-core communication is
quite small. The performance reverts towards the predicted curve after that

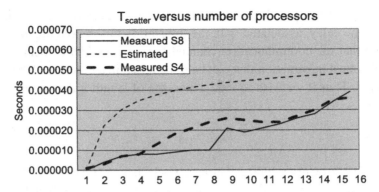

Figure 3-5: Comparison of estimated versus actual scatter operation times $(n = 1200*4*8, \lambda = 3\,\mu s, \beta = 1\,Gbps)$ for 1 to 16 processors.

however, when cores from both blades are involved. Actual performance is better than predicted in this case because of the multicore component.

The graph labeled 'Measured S4' in Figure 3-5 had the first four processors allocated to the first blade, the second four the second blade and so forth. The performance change happens after 4 processors in this graph, as would be predicted.

There is also the potential for actual performance to be worse than predicted. This is due to *process skew* (the binary tree timing of collective communications is disrupted when paired processes have markedly different times) among other factors.

The broadcast and the reduction operations differ from the scatter/gather operations because in each step in the binary tree the message length does not decrease by two as it did before, so we get:

$$T_{reduction} = T_{broadcast} = \sum_{i=1}^{\log p} \left(\lambda + \frac{n}{\beta}\right) = \left(\lambda + \frac{n}{\beta}\right)\log p$$

Figure 3-6 shows the performance of a parallel reduction operation (the graph labeled 'Measured R') on the same cluster as Figure 3-5. The first 8 processors show better than predicted performance, but the second 8 revert to the predicted performance as discussed for scatter/gather.

With a shared communication medium such as Ethernet, broadcast and multicast communications can be much more efficient; however, in that case, point-to-point operations with disjoint sets of senders and receivers

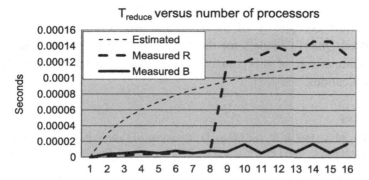

Figure 3-6: Comparison of estimated versus actual reduce and broadcast operation times ($n =$ 1200*4*8, $\lambda = 3\,\mu s$, $\beta = 1\ Gbps$) for 1 to 16 processors.

cannot be done in parallel. The measured performance for the broadcast operation here (the graph labeled 'Measured B') indicates that the switch is likely transitioning from point-to-point operation to shared medium operation for this communication.

3.5. Summary

This chapter has introduced the programming language we will use to build parallel algorithms for robot applications. We have chosen to use the Message Passing Interface (MPI), which is an API for expressing explicit parallelism in C/C++ and Fortran programs (though other MPI language bridges do exist). For the remainder of this book, we will assume familiarity with the C language, and we will introduce MPI commands as they are needed to address problems.

We reviewed three key aspects of MPI:

- The message passing concept.
- The Single Program Multiple Data (SPMD) model.
- Collective communication operations.

MPI implementations are available for many hardware platforms and anybody using Linux may have MPI installed already. We presented an MPI C language template program: a skeleton that can be used as the foundation for building MPI programs. The steps for how to compile and how to run programs that use MPI on a cluster of processors were outlined.

Finally, we introduced the mathematical models we use to evaluate the time complexity of the programs we design. In particular we introduced Quinn's models for the MPI collective communication times. For a comparison of this and other approaches to modeling collective communications, see [12].

References

1. Arbib, M.A., Iberall, T., and Lyons, D., Coordinated Control Programs for Movements of the Hand, in *Hand Function and the Neocortex*, A.W. Goodwin and I. Darian-Smith (Eds), Berlin, Springer-Verlag (1985).
2. Baase, S. and van Gelder, A., *Computer Algorithms*. Addison-Wesley 1999.
3. Chamberlain, B.L., Sung-Eun Choi, Christopher Lewis, E., Calvin Lin, Lawrence Snyder, and Derrick Weathersby, W., ZPL: A machine independent programming language for parallel computers, *IEEE Transactions on Software Engineering*, 6(3):197-211, March 2000.
4. Chapman, B., Jost, G., and van der Pas, R., *Using OpenMP: Portable Shared Memory Parallel Programming*, MIT Press 2008.
5. Cheng, D.Y., A Study of Parallel Programming Languages and Tools. *NASA Ames Research Center Report RND-93-005*, March 1993.
6. Gropp, E., Lusk, E., and Sterling, T., *Beowulf Cluster Computing with Linux*. MIT Press 2003.
7. Hoare, C.A.R., *Communicating Sequential Processes*, Prentice-Hall 1985.
8. Jeannerod, M., Arbib, M.A., Rizzolatti, G., and Sakata, H., Grasping objects: the cortical mechanisms of visuomotor transformation, *Trends Neurosci*. 18:314–320, 1995.
9. Kerridge, J., *Occam Programming: A Practical Approach*. Wiley-Blackwell, 1987.
10. McGraw, J., Skedzielewski, S., Allan, S., Grit, D., Oldehoeft, R., Glauert, J., Dobes, I., and Hohensee, P., SISAL-Streams and Iterations in a Single Assignment Language, Language Reference Manual, version 1. 2. *Technical Report TR M-146, University of California — Lawrence Livermore Laboratory*, March 1985.
11. Message Passing Interface Forum, MPI: A Message-Passing Interface Standard, Version 2.2, 2009.
12. Pjesivac-Grbovic, J., Angskun, T., Bosilca, G., Fagg, G., Gabriel, E. and Dongarra, J., Performance analysis of MPI collective operations, *Cluster Computing*, V10, N2, 2007, pp. 127–143.
13. Quinn, M.J., *Parallel Programming in C with MPI and OpenMP*, McGraw-Hill 2004.
14. Venkataraman, S.T. and Iberall, T., (Eds.) *Dexterous Robot Hands*, New York: Springer Verlag, 1990.
15. Wilkinson, B. and Alen, M., *Parallel Programming*. 2nd Ed., Prentice-Hall 2005.

Chapter 4

Robot Motion

In this chapter, we introduce a cluster implementation of a simple, but illustrative, computational problem in robotics — calculating the final and intermediate locations of a robot, or a team of robots, when a sequence of motions is executed. This fairly straightforward computation is used to introduce the steps of *partitioning*, *program design* and *analysis of performance* used throughout the volume.

There are many ways to categorize robot platforms. One categorization bears a resemblance to an informal division of all life into animals and plants; it categorizes platforms based on whether the robot can move around its environment or whether it occupies a fixed spot. Each kind of platform can be very effective in its own task niche:

- *Stationary robots* are common in manufacturing [2]. They typically work on product subassemblies which are conveyed to and from the robot's *workspace* (the volume of space that the robot can usefully reach) by some other transportation system. Historically, the industrial manipulator was the first robot platform to have a strong economic justification [1].
- *Mobile robots* are robot platforms that are equipped with a locomotion functionality that allows them to move themselves around their environment [3, 4]. aerial [5], climbing [7] and underwater [6] robots have all established themselves as effective classes of mobile robot. In this text we confine ourselves to ground-based robots.

Figure 4-1:   Pioneer 3-DX (left), Pioneer 3-AT (center), Sony AIBO (right).

The two principal methods of ground robot locomotion are *legged locomotion* — where a collection of limbs moves the robot by stepping — and *wheeled locomotion* — where the robot platform rolls on wheels. Of course, many different configurations of each of these exist and certainly other kinds of ground-based locomotion are possible (e.g., snake-like motion, hopping, etc; see Chapter 7 in [4] for a good summary).

Figure 4-1 shows a collection of robots. The leftmost image is of a two-wheeled *Pioneer 3-DX* robot built by *Adept Mobilerobots*. It is controlled by the laptop computer mounted on top and is equipped with stereovision, video and sonar sensors (the light-colored disks mounted on the side of the robot just under the top plate are the sonar sensors). A four-wheeled *Pioneer 3-AT* is shown in the middle image equipped with a stereovision camera mounted on a pan-tilt (PT) base as well as with sonar sensors. The final image in Figure 4-1 is an *AIBO*®, a four-legged 'dog' robot created by *Sony* as a personal entertainment robot.

There are many different designs possible for the wheel mechanism in a wheeled robot. The *Pioneer 3-DX* in Figure 4-1 has two, independently-driven wheels mounted on the side of the robot, and an undriven, caster wheel mounted under the rear of the robot for balance. Let's consider how the robot will move when the wheel motors are driven in three different ways:

(1) If both wheels motors are instructed to go forward at the same rotational velocity, then the robot will move straight forward[a].

---

[a]This is a simplification of course: if one of the wheels slipped, or struck a rock or other object, the robot would no longer move straight forward.

(2) If the wheels are driven in opposite directions with the same rotational velocity, then the platform will rotate in place, not changing its position, but changing the direction in which the front of the robot points, that is, changing the *orientation* of the robot.

(3) If the wheels are driven forward, but at different constant rotational velocities, then the robot will move forward, and follow a curved path.

Note that there is no way for the robot to move directly sidewards. A constraint such as this on what motions can be directly commanded, where the change in position is not independent of the change in orientation, is referred to as a *nonholonomic motionconstraint*.

4.1. Motion of a Mobile Robot in Two Dimensions

A mobile robot will have some physical dimensions that dictate what volume of space it occupies. As you might imagine, these dimensions can change dramatically depending on the kind of robot: An *Adept MobileRobots Pioneer 3-AT* is typically *501×493×277* mm, whereas the *Adept MobileRobots Seekur*, a much bigger platform, is *1.4×1.3×1.1* m. In contrast, a *K-Team Khepera®III*, a much smaller platform, is *130×130×70* mm. The size of a robot does not typically change as the robot moves around its environment. Because of this invariance, a common simplification is to consider the location of the robot to be the location of the center point (or some other significant, known point) of the robot. We can measure and record the location of this point, the location of the robot, by superimposing an imaginary fixed grid, a Cartesian coordinate system, on the ground around the robot. We will refer to this as the *global coordinate frame*. The position of the robot can be determined by measuring the distance of its center along the x and y axes of this global coordinate frame (see Figure 4-2).

Most robot platforms have a specific orientation — a front — and it is often important to know which way this is facing. There may be sensors, or a robot arm, mounted on the front, or some other fixed location on the robot platform. Knowing what direction the front of the robot faces is crucial to being able to move the sensors to view some target object, or to bring the arm to a location where it can manipulate or interact with a target object. Thus, in addition to knowing the coordinates of the center

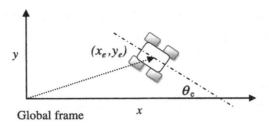

Figure 4-2:   Robot location and orientation in global coordinate frame.

of the robot, it's important to know the orientation of the robot platform
with respect to the global coordinate frame. The *principal axis* of the robot
is a line through the robot, from front to back, and typically centered
with respect to the sides of the robot. We use the angle between this
line and the x-axis as the orientation of the robot. In summary, we can
capture the pose of the robot as the location of its center and orientation
of its principle axis, (x, y, θ) where x and y are distance measurements
in mm and θ is an angular measurement in *radians*, unless otherwise
noted.

4.2.   Calculation of Location by Dead-Reckoning

One of the easiest ways to navigate a nonholonomic mobile robot, such
as the *Pioneer 3-AT* shown in Figure 4-1, around its environment is repeat
the following two-step scheme as many times as needed: First, change the
orientation of the robot until it faces in a desired direction; and second,
move the robot forward a desired amount in that direction. A natural
question to ask is, after a single two-step sequence such as this, what
pose (x_e, y_e, θ_e) will the robot be at? Let us suppose that the robot turns
through some angle θ and moves a distance r_d forward, and that it starts
at (x_s, y_s, θ_s) The result of these motions can be easily expressed by the
dead-reckoning expression:

$$\theta_e = \theta_s + \theta$$
$$\begin{bmatrix} x_e \\ y_e \end{bmatrix} = \begin{bmatrix} x_s \\ y_s \end{bmatrix} + r \begin{bmatrix} \cos \theta_e \\ \sin \theta_e \end{bmatrix} \tag{4.1}$$

In fact, if the robot makes a sequence of such straight-line motions $\{(\theta_i, r_i) | i \in \{1, \ldots, n\}\}$ then the final position will be the sum:

$$\Theta_i = \theta_s + \sum_{j=1}^{i} \theta_j$$

$$\begin{bmatrix} x_e \\ y_e \end{bmatrix} = \begin{bmatrix} x_s \\ y_s \end{bmatrix} + r_i \sum_{i=1}^{n} \begin{bmatrix} \cos \Theta_i \\ \sin \Theta_i \end{bmatrix} \tag{4.2}$$

If there are a large number of motions, or if we are calculating locations for a large number of robots, it may be more effective (i.e., faster) to carry out this calculation in parallel on a cluster of computers than on a single computer. We can calculate the solution to this dead-reckoning problem on the cluster by allocating some contiguous portion of the sum to each processor and then summing the results from each processor. This example is straightforward enough that it will allow us to introduce the basics of cluster programming in MPI.

We will assume that Θ_i has been calculated, and each processor calculates its portions of $r_i \sum_{i=1}^{n} \begin{bmatrix} \cos \Theta_i \\ \sin \Theta_i \end{bmatrix}$ before returning its results for the final summation. We must first address the question of how to distribute the computation among the processors in the cluster. We call this step *partitioning*. In this example, as in many we will look at, the processors are all doing basically the same operations, and the partition issue boils down to how to assign the motion data to processors in the cluster of p processors so that the results are calculated as quickly as possible.

4.2.1. *Partitioning: Block Data Decomposition*

In [8] Quinn introduces *block data decomposition*: dividing an array into p roughly equal-sized blocks that are to be distributed over the p processors that are carrying out the parallel calculation. Each processor is provided with roughly the same amount of data, and thus will take roughly the same amount of time to finish its portion of the work load. In this case, we say that the work load is *balanced*. The parallel computation time for this situation is the time of the slowest processor. With a balanced work load, this time is the same for every processor, and no processor inefficiently idles while waiting for its slower neighbor to finish.

```
#define BLOCK_LOW(id,p,n) ((id) * (n) / (p))
#define BLOCK_HIGH(id,p,n) (BLOCK_LOW((id)+1,p,n) – 1 )
#define BLOCK_SIZE(id,p,n) (BLOCK_HIGH(id,p,n) – BLOCK_LOW(id,p,n) + 1)
#define BLOCK_OWNER(index,p,n) ( ( (p)*( (index) + 1 ) – 1 ) / (n))
```

Figure 4-3:   Quinn's block allocation macros [8].

In the case where p divides n evenly, each processor is allocated an identical amount $\frac{n}{p}$ of the array. If p does not divide n evenly, then we have to deal with issue of the remainder. The operation $\lceil v \rceil$ is the *ceiling* of v, defined as the next largest integer value to the number v. The operation $\lfloor v \rfloor$ is the *floor* of v, defined as the next smallest integer value to the number v. A simple block decomposition approach would be to allocate $\lceil \frac{n}{p} \rceil$ to every processor except the last. We then give the remainder, which is $n - (p - 1) \lceil \frac{n}{p} \rceil$, to the last processor. However, the remainder might be zero or quite small, resulting in an *unbalanced work load* and the inefficiency of an underused processor. A better approach is to balance the work assignments for each processor in the cluster by giving it either $\lceil \frac{n}{p} \rceil$ or $\lfloor \frac{n}{p} \rfloor$ elements of the array. Quinn introduces a scheme that has relatively simple index calculations and also spreads the distribution of $\lceil \frac{n}{p} \rceil$ and $\lfloor \frac{n}{p} \rfloor$ sized blocks evenly over the values of the processor index (rather than collecting all the $\lfloor \frac{n}{p} \rfloor$ sized blocks together in the low-number indices, for example). The scheme is captured in the C macro definitions in Figure 4-3.

The BLOCK_LOW and BLOCK_HIGH macros will return the data index of the first and last data items, respectively, allocated by the block data decomposition to processor id, where there are p processors and n data items in total. BLOCK_SIZE returns the number of data items allocated to processor id, while BLOCK_OWNER will produce the processor id that has been allocated to process data item index. We will make widespread use of these macros in this text.

4.2.2.   *Program Design*

With this data allocation scheme we can write the MPI code to partition the dead-reckoning position calculations on a cluster. The first draft at a

```
MPI_Init(&argc,&argv);
MPI_Comm_rank(MPI_COMM_WORLD,&id);
MPI_Comm_size(MPI_COMM_WORLD,&p);
x =  y = 0;
for (i = BLOCK_LOW(id,p,n); i<=BLOCK_HIGH(id,p,n); i++) {
        x += r[ i ] * cos( THETA[ i ] );
        y += r[ i ] * sin( THETA[ i ] );
}
```

Figure 4-4:   Sketch of dead-reckoning MPI code.

solution for this program is shown in Figure 4-4. The initial three MPI function calls in the code section will be common to all our programs; they establish the processor id (rank), and the total number of processors, p. The for-loop carries out the calculation of the intermediate locations (x, y) for values of i within the block of movement data allocated to processor id. When the loop finishes, the variables x and y hold the final pose for this subsequence of motions.

This draft code illustrates the main processing steps for dead-reckoning, but it has some serious flaws. It assumes that the motion arrays r[ ], containing the forward distances, and THETA[ ], containing the cumulative orientation change, are available to all processors. This will not be the case; the motions must be transmitted to the processors. It also does not complete the final step of adding all the subsequence poses and the start pose. We will address each issue in turn.

We could simply send the motion arrays r[ ] and THETA[ ] to all processors using an MPI broadcast communication command. This would however waste communication time, since each processor will only use the portion of the motion arrays allocated to it by the BLOCK_LOW and BLOCK_HIGH macros. Instead, we use a scatter operation, MPI_Scatter, to distribute the arrays in data blocks to each processor (see Figure 4-5). This reduces communication time by distributing to each processor only the data that it needs. Let us assume for convenience that n is a multiple of p so that each processor can be allocated the same block size, n/p.

The scatter command is a collective communication operation — that is, all the processors in the communicator (MPI_COMM_WORLD in general) participate in the operation. The following is the function prototype for MPI_Scatter:

```
int MPI_Scatter (    void            *sendbuf,
                     int             sendcnt,
                     MPI_Datatype    sendtype,
                     void            *recvbuf,
                     int             recvcnt,
                     MPI_Datatype    recvtype,
                     int             root,
                     MPI_Comm        comm          )
```

The **sendbuf** parameter is the address of the array (buffer) being scattered. The **sendcnt** parameter contains the size of the block to be sent to each processor and **sendtype** is the type of data elements in the block. All the MPI commands introduced in the book are summarized in Appendix IV for convenience. Appendix II summarizes the list of MPI data types (**MPI_Datatype** in the prototype above).

Some processor has to read in the motion arrays from disk or by communicating with the robot(s) — the *root processor*. Typically the root processor is the processor with index 0. The send parameters are only important for this processor. The receive parameters are important for the root and all other processors, as they indicate where the scattered data will be put. The parameter **recvbuf** is the address of the array (buffer) into which the scatter elements are to be placed, where **recvcnt** is the size of the block and **recvtype** the type of each element. Finally **root** is the rank of the root processor.

There are two arrays that need to be scattered: r[ ] and THETA[ ]. The first step is to calculate the size of the blocks to be distributed. Since we assume that n is a multiple of p, this size is n/p. We can call MPI_Scatter as follows.

```
MPI_Scatter( &r[ 0 ], n/p, MPI_FLOAT,
             &r[ BLOCK_LOW( id, p, n) ], n/p, MPI_FLOAT,
             0, MPI_COMM_WORLD);
```

Each processor will receive its n/p portion of the array into &r[BLOCK_LOW(id,p,n)]. This MPI command will distribute the forward distance array r[ ]. We can write a similar MPI command for the cumulative turn angle array THETA[ ].

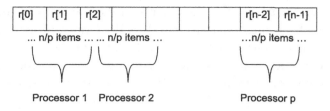

Figure 4-5: Scattering of one array.

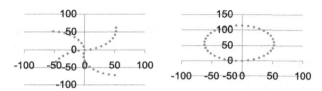

Figure 4-6: Robot moving in a circle of 36 steps; four 0-based calculations of 9 steps in order anticlockwise (left); adding the end point of the previous 9 step sequence to each of the points in the next 9 step sequence reproduces the original circle (right).

To calculate the values of THETA correctly, each processor needs to start its subtotal not at zero, as we do in Figure 4-4, but at whatever the immediately previous block of moves produced. That is, the subtotal results of the block of motion calculations on each of the processors need to be added together. This is more than simply a reverse of the scattering operation since the subtotals need not just to be transmitted to the root processor, but to be added together to produce the correct result (as illustrated in Figure 4-6).

The collective communication function MPI_Reduce will perform a *reduction operation* on data from every processor and place the result of the operation on the root processor. A reduction operation is a binary associative operation (these include sum and product, but also maximum, minimum and logical operations; see Appendix III for a full list of MPI's built-in reduction operations). For the location calculations, we need to reduce all the subtotals, the x and y variables, using a summation reduction. The prototype for MPI_Reduce is as follows.

```
int MPI_Reduce (     void              *sendbuf,
                     void              *recvbuf,
                     int               count,
                     MPI_Datatype      datatype,
                     MPI_Op            op,
                     int               root,
                     MPI_Comm          comm     )
```

The parameter sendbuf is the address of the subtotal and recvbuf is the address of the final reduced total. If a list (array) of items is being separately reduced, then count can be used to indicate the size of the array. The reduction operation is specified by the value of op. The remaining parameters are similar to those for MPI_Scatter. To reduce the value of x we call:

MPI_Reduce(&x, &xE, 1, MPI_FLOAT, MPI_SUM, 0, MPI_COMM_WORLD)

where x is the local subtotal and where xE holds the reduced total on the root processor, the processor with rank 0. Once the root processor gets the values of xE and yE, it concludes the calculation by adding in the start positions. This completes all the steps we need to do the cluster calculation of final position, and the entire code is shown in Figure 4-7.

4.2.3. *Analysis*

To understand the benefits and trade-offs of using a cluster to calculate location, we should compare the sequential and parallel computing costs. The main calculation loop for dead-reckoning requires two multiplications and two additions repeated n times and then 2 final additions. If the addition time is a seconds and the multiplication time is seconds then the overall loop takes

$$T_s = 2n(a + b) + 2a.$$

The parallel time consists of a calculation (computation) time T_{cal} and a communication time T_{com}. The main loop calculation time is reduced to $T_{cal} = 2n(a + b)/p + 2a$ by parallelism since each processor only handles n/p items, but the root processor still needs to do the final two additions.

The parallel communication time includes the scattering of the two arrays and then the reduction of the two results. Each collective communication among p processors requires $\log p$ steps. The message size for each scatter operation is n/p. As we saw in Chapter 3, the total time for each scattering is therefore $\lambda \log p + n(p-1)/p\beta$ where λ is the communication latency and β the communication bandwidth.

```
MPI_Init(&argc,&argv);
MPI_Comm_rank(MPI_COMM_WORLD,&id);
MPI_Comm_size(MPI_COMM_WORLD,&p);

if (id==root) {
            /* read in motion arrays r[] and theta[] and start positions
               xS, yS and thetas calculate cumulative angle array THETA[] */
}

/* Scatter the motion arrays to each processor in the cluster */
MPI_Scatter( &r[0], n/p, MPI_FLOAT,
             &r [BLOCK_LOW(id,p,n)], n/p, MPI_FLOAT,
             0, MPI_COMM_WORLD );
MPI_Scatter( &THETA[0], n/p, MPI_FLOAT,
             &THETA [BLOCK_LOW(id,p,n)], n/p, MPI_FLOAT,
             0, MPI_COMM_WORLD );

/* each processor calculates its section of the motion */
x = y = 0;
for (i = BLOCK_LOW(id,p,n); i<=BLOCK_HIGH(id,p,n); i++) {
            x += r[ i ] * cos( THETA[ i ] );
            y += r[ i ] * sin( THETA[ i ] );
}

/* and the sub totals are reduced in xE and yE*/
MPI_Reduce(&x, &xE, 1, MPI_FLOAT, MPI_SUM, 0,
           MPI_COMM_WORLD);
MPI_Reduce(&y, &yE, 1, MPI_FLOAT, MPI_SUM, 0,
           MPI_COMM_WORLD);

If (id==root) { /* Finally the root adds in the start position */
            xE += xS;
            yE += yS;
            /* and any other output related work will go here*/
}
```

Figure 4-7: MPI Program for cluster calculation of final robot location.

The size of the message passed in the reduction operation is a constant independent of n, so the reduction time is essentially $\lambda \log p$. The total time spent in communication is therefore:

$$T_{com} = 2(\lambda \log p + n(p-1)/p\beta) + \lambda \log p$$

and the overall parallel execution time is:

$$T_p = T_{cal} + T_{com}$$
$$= \left[\frac{2n(a+b)}{p} + 2a \right] + \left[3\lambda \log p + \frac{2n(p-1)}{p\beta} \right]$$

The expressions for the serial and parallel execution times can be used to estimate the benefit of partitioning the computation over p processors. See Figure 4-9 for performance graphs of these times, but discussion is postponed until after the next section.

4.3.   Dead-Reckoning with Intermediate Results

Mapping is the process of using the information about the robot's position and the information from its sensors to construct a representation of its environment — a map. Localization is the process of using information from the robot's sensors to determine where the robot is on its map. A first estimate of where the robot is on a map can be obtained by using the dead-reckoning calculations from the previous section. Due to many factors, a robot may not end up in the position predicted by dead-reckoning — a wheel may slip during motion, the slope of the terrain may slow down or speed up the vehicle, etc. However, by carefully comparing the sensor readings that a robot receives at each location with what might be expected from its sensors given the map, the robot can 'fine-tune' its position estimate. We will discuss the implementation of some approaches to localization later. However, a first estimate at localization is the sequence of locations calculated purely by dead-reckoning. Note that Eq. (4.2), the basis of our dead-reckoning implementation, produces only the final location after n moves and not the $n-1$ intermediate locations in addition to the final location.

4.3.1. *Partitioning*

It seems as if it will be straightforward to modify our approach to produce the intermediate locations, after all, they are calculated, just not stored. We can modify the main loop of our dead-reckoning program in Figure 4-7 to store the intermediate positions to an array of x and y values and transmit this entire array back to the root. Therein lies a problem: we did not transmit back the subtotals, we carried out a parallel reduction to sum the subtotals to produce the final location. Each processor calculates its section of motion as if the robot started at $(0, 0)$ but with the correct orientation from the THETA array.

Summing these does indeed produce the final location, but that won't work for intermediate locations. We can either design the array of intermediate locations so that it can be effectively reduced or we can still send subtotals but leave it to the root processor to add in all the offsets in a loop. The reduction is the more efficient approach since it accomplishes the operations in $O(\log p)$ as opposed to $O(p)$ for the root processor's addition loop.

We set up the array of intermediate locations on each processor so that when each element of the array is summed during reduction it will produce the correct intermediate position. An example illustrating this is shown in Figure 4-8. When a processor finishes its block of calculations, it continues and fills the remainder of the entire array with the value of the last location it calculated (the subtotal from the previous example). If the array for each

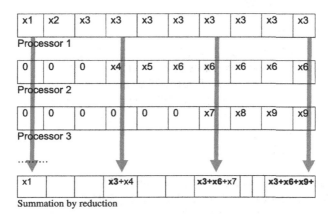

Figure 4-8: Design of intermediate result arrays to facilitate reduction, for the case $n/p = 3$, and only showing the first three processors and the reduced array.

processor was written as a row of a matrix, the resulting matrix would have a stepped upper diagonal form (Figure 4-8). When corresponding elements (columns in the figure) are summed, the repeated elements correctly offset the $(0, 0)$ based location calculations of processors summing motions that had occurred 'later' in the motion sequence. This approach also gives us an elegant way to take the start position of the robot into account: The root processor adds the start positions to its portion of the calculations at the start rather than at the end of its loop. This incurs the same computational cost as before.

4.3.2. *Program Design*

In summary, there are two principal changes to modify the dead reckoning code to produce intermediate positions: Each processor now executes $2n(1 - (id + 1)/p)$ extra assignments to fill in the offsets for later motion, and the reduction statements now operate on the $2n$ items in the xL and yL lists of intermediate positions. The code is as follows.

```
/* each processor calculates its section of the motion and offset for later
motion */
if (p==0) { /* root takes start position into account */
        x = xS; y = yS;
}
else { /* other processors produce (0,0) based motion */
        x = y = 0;
}
for (i = BLOCK_LOW(id,p,n); i<n; i++) {
        if ( i <= BLOCK_HIGH(id,p,n) ) {
                /* motion done by this processor */
                x += r[ i ] * cos( THETA[ i ] );
                y += r[ i ] * sin( THETA[ i ] );
        }
        xL[i] = x;
        yL[i] = y;
}
/* and the sub totals are reduced in xLE and yLE */
MPI_Reduce(&xL, &xLE, n, MPI_FLOAT, MPI_SUM, 0,
                MPI_COMM_WORLD);
MPI_Reduce(&yL, &yLE, n, MPI_FLOAT, MPI_SUM, 0,
                MPI_COMM_WORLD);
```

4.3.3. *Analysis*

While the change to the parallel calculation time T_{cal} is negligible, the modification of the reduction operation means we need to add in a bandwidth-related term for the message length. The reduction will still take $\log p$ steps in each case, but the message length is now the full n array elements. Following Chapter 3, this will take $(\lambda + n/\beta) \log p$ for reduction of x and the same for reduction of y. The communication time becomes

$$
\begin{aligned}
T_{com} &= 2T_{red} + 2T_{scat} \\
&= 2\left[2\lambda + \frac{n}{\beta}\right] \log p + \frac{2n(p-1)}{p\beta}
\end{aligned}
$$

Figure 4-9(a) shows the total execution times for the final-position and intermediate-position versions of parallel dead-reckoning versus the serial time for final-position, calculated using the expressions developed here and in the previous section. While the parallel intermediate position has a speed advantage initially, the cost of the parallel reduction causes the total execution time to slowly rise as the number of processors is increased, and for $n = 4(bytes/float)*12000(floats)*8(bits per byte)=384000$ bits in Figure 4-9(a) the point at which the parallel execution time equals the serial time is about 5 processors. With increasing n, this point moves further out along the horizontal axis, but the shape of the three curves remains the same.

Figure 4-9(b) and (c) shows graphs of the measured execution times for the dead-reckoning code developed in this chapter executed on a cluster of dual-processor, quad-core Intel Xeon® processors connected with a gigabit HP ProCurve® switch. Two dual quad-core blades were used for the test. In each case, the execution of the program on the initial one up to eight processors all show much lower than predicted execution timing. This is reasonable, since all eight processors are cores on the same blade server. Once the processors of the second blade are deployed (when the code is executed on nine and more processors) the execution time quickly rises to the predicted values from Figure 4-9(a).

4.4. **Dead-Reckoning for a Team of Robots**

A group or team of robots operating cooperatively can accomplish tasks that single robots cannot [4]. For example: exploring dangerous terrain where the failure of a single robot would mean the failure of the mission; carrying out

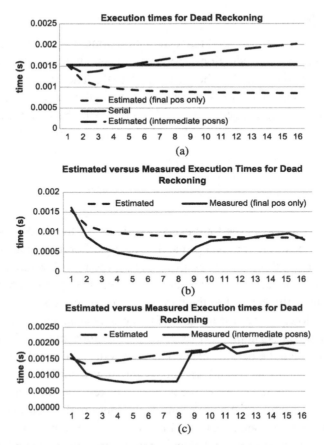

Figure 4-9: (a) Graph of estimated total execution times for parallel final position, serial final position, and parallel intermediate position for 1 to 16 processors ($n = 12000$, $a = b = 7ns$, $\lambda = 3\mu s$, $\beta = 1$ *Gbps*). (b) Graph of execution times on HPC cluster for parallel final position only. (c) Graph of execution times on HPC cluster for parallel intermediate position.

activities that require more than one robot such as pushing or lifting cooperatively, or activities that need the sensing, measurement or manipulation resources from differently configured platforms; carrying out tasks more quickly than a single robot can, e.g., exploring or mapping an area.

We will extend the dead-reckoning problem to a team of robots as follows. Let us consider a team of m robots. For each robot $j \in \{1, \ldots, m\}$ we have the forward distance moved in each of its $i \in \{1, \ldots, n\}$ steps in the matrix $[r_{ji}]$. We also have the cumulative angle matrix $[\Theta_{ji}]$ defined by the angular summation from the start θ_{sj} angle for robot j to the ith move. We can define the intermediate position after the ith move for robot j by an extension of Eq. (4.2):

$$\Theta_{ji} = \theta_{sj} + \sum_{k=1}^{i} \theta_{jk}$$

$$\begin{bmatrix} x_{ji} \\ y_{ji} \end{bmatrix} = \begin{bmatrix} x_{sj} \\ y_{sj} \end{bmatrix} + r_{jk} \sum_{k=1}^{i} \begin{bmatrix} \cos \Theta_{jk} \\ \sin \Theta_{jk} \end{bmatrix} \tag{4.3}$$

The objective of the dead-reckoning problem for a team of robots is to produce the matrix of $p_{ji} = \begin{bmatrix} x_{ji} \\ y_{ji} \end{bmatrix}$ values.

4.4.1. *Partitioning*

For the single robot case, there were no alternate partition approaches to be considered, other than the issue of balancing computation on all processors. However, in general from now on, there will be more than one approach to partitioning the data and computation over the cluster. We use the models and metrics introduced at the end of Chapter 3 to select between the different partitioning approaches.

Let us consider two avenues open to us for partitioning the multiple robot dead-reckoning calculations on a cluster:

(1) We can distribute the calculations by column (the i index), as we did for the single robot case. In that case, each processor will calculate some (n/p) of the motions for all robots — the computation and communication times will simply be m times the single robot case.

(2) We can distribute the calculations by row (the j index). In that case, each processor will calculate all of the motions for some (m/p) robots. While the computation and scattering are the same as for distribution by row, because each processor calculates all the summations for its assigned robots, no parallel reduction is necessary. Instead, all the intermediate results need only to be put 'back together' into the $p_{ji} = \begin{bmatrix} x_{ji} \\ y_{ji} \end{bmatrix}$ matrix on the root processor. This operation — the reverse of a scatter- is called a *gather*. This should represent a savings in communication time, so we adopt this as our approach.

4.4.2. *Program Design*

Let us assume that we have stored the two motion arrays such that r[j][i] and THETA[j][i] refer to the ith motion step for the jth robot. In C, two dimensional arrays are stored in *row-major order*. That is, row 0 occupies memory sequentially and is followed immediately by row 1 and so on. Hence scattering the first m/p rows (each of length n) to processor 0, the second m/p rows to processor 1, and so forth, requires little effort. For example, for the forward motion matrix, we will have:

```
MPI_Scatter( &r[0][0], n*m/p, MPI_FLOAT,
                &r[BLOCK_LOW(id,p,m)][0], n*m/p, MPI_FLOAT,
                0, MPI_COMM_WORLD );
```

(again making the assumption that p divides m evenly). The calculation consists of two nested for-loops: The first spans the m/p robots assigned to the processor and the second spans all the motion steps for this robot. However, the start locations for the robots assigned to each processor need to be added in explicitly this time. Therefore, we need to scatter the start positions, xS[j] and yS[j], to each processor; an additional two scatter operations with a combined message size of $2m/p$.

```
for (j = BLOCK_LOW(id,p,m); j<BLOCK_HIGH(id,p,m); j++) {
       x = xS[j];y = yS[j];
       for (i=0; i<n; i++) {/* all motion for this rbt done by this proc*/
                xL[ j ][ i ] = x += r[ j ][ i ] * cos( THETA[ j ][ i ]);
                yL[ j ][ i ] = y += r[ j ][ i ] * sin( THETA[ j ][ i ]);
       }
}
```

The final step is the gather collective communication operation. The function prototype for MPI_Gather is

```
int MPI_Gather(      void              sendbuf,
                     int               sendcnt,
                     MPI_Datatype      sendtype,
                     void              *recvbuf,
                     int               recvcount,
                     MPI_Datatype      recvtype,
                     int               root,
                     MPI_Comm          comm     ).
```

The parameters are very similar to those for scatter: The first three parameters give the address of the buffer on the sending processor, its size and its type. The second three give the address, size and type on the receiving (root) processor. The final two are the index of the root processor and the communicator.

The m/p rows of xL and yL calculated by each processor need to be reassembled, gathered, into a single matrix containing all the calculations on the root processor. For the x component, this is achieved with the following MPI command:

```
MPI_Gather( &xL[0][0], n*m/p, MPI_FLOAT,
            &xLE[BLOCK_LOW(id,p,m)][0], n*m/p, MPI_FLOAT,
            0, MPI_COMM_WORLD );
```

4.4.3. *Analysis*

Let us consider the two approaches in more detail.

(1) *We can distribute the calculations by column*, as we did for the single robot case. In that case, each processor will calculate some of the motions (n/p) for all robots. The parallel computation time is simply m times the single robot case: $m(n/p)$ calculations of the intermediate positions, $T_{cal} = m(2n(a + b)/p + 2a)$. The scattering operation will need to scatter $2mn/p$ elements from the motion matrices r[ ] and THETA[ ] and the parallel reduction will require a message size per processor of mn, giving a total time therefore:

$$T_p = 2m \left[\frac{n(a + b)}{p} + a \right] + 2m \left[2\lambda + \frac{n}{\beta} \right] \log p + \frac{2mn(p - 1)}{p\beta}$$

(2) *We can distribute the calculations by row*. In that case, each processor will calculate all of the motions for some (m/p) robots. The parallel computation remains the same: $n(m/p)$ calculations of the intermediate position. The message size for scattering becomes $2(n + 1)m/p$. However, no parallel reduction phase is called for. The data that need to be gathered are the x_{ji} and y_{ji} components for the n steps of the

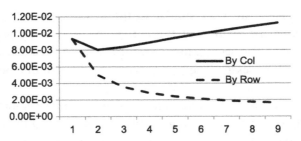

Figure 4-10: Graph of execution time versus number of processors for distribution by row and distribution by column *(m =100, n =400, a = b =7 ns, λ =3 μ s, β =1 Gbps).*

m/p robots assigned to each processor: a message length of $2mn/p$. In that case, the total parallel time is:

$$T_p = 2m \left[\frac{n(a+b)}{p} + a \right] + \left[4\lambda \log p + \frac{2m(2n+1)(p-1)}{p\beta} \right]$$

Figure 4-10 shows a graph of T_p versus number of processors for both approaches; the row decomposition approach has the better performance.

4.4.4. *Local and Global Buffers*

In the previous examples, the full matrices for r, THETA, xL and yL are replicated on every processor. This makes the algorithm easier to read. However, it is wasteful of memory. A more efficient approach would be to dynamically allocate on each processor only the amount of memory that we need. Recall that in C, 2D arrays are stored row sequential (row-major form). A two dimensional $m \times n$ array consists of an $m \times n$ block of memory containing the data and an $m \times 1$ row-address array containing the start of each row address. The following code (from [8]) will dynamically allocate such an array:

```
ElementType **arrayData, *arrayName;

arrayData = (ElementType *)malloc(n*m*sizeof(ElementType));
for (i=0; i<m; i++) arrayname[i] = &arrayData[i*n];
```

On each of the processors other than the root, we only need enough storage to hold the $(m/p) \times n$ block of data for the r, THETA, xL and yL arrays — that is a 2D array of size $(m/p) \times n$. These local arrays — rloc,

THETAloc, xLloc, ylLoc — can be allocated as explained above. All that remains is to modify the scatter and gather commands to use these arrays, and to modify the use of the loop indices in the calculation portion of the code. This modified code is shown below (Note: &r[0][0] is the same as r, but the longer form is used here to make the visual connection with the prior examples.)

```
/* Allocate local arrays of dim (m/p)×n: rloc, THETAloc, xLloc, yLoc */
/* scatter r,THETA into the local arrays rloc, THETAloc */
    MPI_Scatter( &xS[0], m/p, MPI_FLOAT,
                 &xSloc, m/p, MPI_FLOAT,
                 0, MPI_COMM_WORLD );
    MPI_Scatter( &yS[0], m/p, MPI_FLOAT,
                 &ySloc, m/p, MPI_FLOAT,
                 0, MPI_COMM_WORLD );
    MPI_Scatter( &r[0][0], n*m/p, MPI_FLOAT,
                 &rloc[0][0], n*m/p, MPI_FLOAT,
                 0, MPI_COMM_WORLD );
    MPI_Scatter( &THETA[0][0], n*m/p, MPI_FLOAT,
                 &THETAloc[0][0], n*m/p, MPI_FLOAT,
                 0,MPI_COMM_WORLD );
    /* calculate motions using the local arrays) */
    for (j = 0; j< BLOCK_SIZE(id,p,m); j++) {
        x = xSloc[ j ]; y = ySloc[ j ];
        for (i=0; i<n; i++) { /* all motion for this robot done by this prc */
            xLloc[ j ][ i ] =
                x += rloc[ j ][ i ] * cos( THETAloc[ j ][ i ] );
            yLloc[ j ][ i ] =
                y += rloc[ j ][ i ] * sin( THETAloc[ j ][ i ] );
        }
    }
    /* gather the result from local arrays back on the root processor */
    MPI_Gather( xLloc, n*m/p, MPI_FLOAT,
                &xLE[BLOCK_LOW(id,p,m)][0], n*m/p, MPI_FLOAT,
                0, MPI_COMM_WORLD );
    MPI_Gather( yLloc, n*m/p, MPI_FLOAT,
                &yLE[BLOCK_LOW(id,p,m)][0], n*m/p, MPI_FLOAT,
                0, MPI_COMM_WORLD );
```

4.5. Summary

In this chapter, we introduced some of the basic concepts, notation and terminology in mobile robotics. We introduced some computational problems related to calculating robot motion, and we looked at three examples:

(1) Calculation of the end location of a robot motion after a sequence of moves — *dead-reckoning*.
(2) Calculation of the intermediate as well as final dead-reckoning locations.
(3) Dead-reckoning for a team of robots.

The first step in understanding how a computational problem can be implemented on a cluster is to explore the partition of the computation and data on the processors of the cluster. We used the communication time expressions from Chapter 3 to model the parallel execution time of our proposed algorithm and to compare it to the serial time. Quinn's block decomposition macros were introduced as a means to partition data over a cluster in a balanced fashion, and the MPI_Scatter command was presented as a way to carry out this partitioning. The MPI_Reduce command was introduced as a way to carry out a parallel, logarithmic time summation.

The parallel implementation of dead-reckoning with intermediate results showed us the potential negative effect of communication overhead — the performance of this algorithm eventually becomes crippled by its communications costs.

Extending the dead-reckoning problem to a team of robots allowed us to introduce the important concept of evaluating various partitioning approaches before committing to an algorithm. In this example, we looked at a row versus a column partitioning of the problem and could show that the row had the better performance.

Finally, the savings introduced by using a local array, rather than reusing the 'global' array, to hold scattered data was explored. To leverage these memory savings, the local array has to be dynamic.

References

1. Engelberger, J.F., *Robotics in Practice*, MIT Press 1989.
2. Nof, S.Y., *Handbook of Industrial Robotics*, Wiley 1999.

3. Arkin, R.C., *Behavior-Based Robotics*, MIT Press 1998.
4. Bekey, G.A., *Autonomous Robots*, MIT Press 2005.
5. Lewis, M.A., Fagg, A.H., and Bekey, G.A., The USC Autonomous Flying Vehicle: an Experiment in Real-Time Behavior-Based Control, *Proceedings of the IEEE Conference on Robotics and Automation*, May, 1993, Atlanta, Georgia.
6. Yuh, J., Design and Control of Autonomous Underwater Robots: A Survey, *Autonomous Robots*, 8(1), January 2000.
7. Asbeck, A.T., Kim, S., Cutkosky, M.R., Provancher, W.R., and Lanzetta, M., Scaling hard vertical surfaces with compliant microspine arrays, *Proceedings, Robotics Science and Systems*, Cambridge MA, 2005.
8. Quinn, M.J., *Parallel Programming in C with MPI and OpenMP*, McGraw-Hill 2004.
9. Thrun, S., *Robotic Mapping: A Survey*. In G. Lakemeyer and B. Nebel, editors, *Exploring Artificial Intelligence in the New Millenium*. Morgan Kaufmann, 2002.

Chapter 5

Sensors

A mobile robot platform which has no ability to sense its environment suffers two disadvantages:

(a) It cannot validate its locomotion actions — and so may end up in a different location than it expected, and
(b) should the environment in which the robot finds itself differ from its preprogrammed knowledge, it may, at the very least, be unable to achieve its task objectives.

For these reasons, most robot platforms are equipped with sensors. While animals tend to have a lot of sensors [1] to help move around effectively, robot platforms tend to have fewer [2]. Sensors are typically divided into *exterioceptors*, exterior measuring sensors, and *proprioceptors*, interior measuring sensors. Interior and exterior in the robot case refer to the measurement of properties of the robot platform, and properties of the robot's environment, respectively.

Sensors that measure internal state include those that measure:

- the amount of wheel rotation (e.g., an odometer) or leg movement;
- the inclination of the platform relative to the gravity vector (e.g., an inclinometer);
- the orientation of the platform (e.g., compass heading);
- the forces and torques exerted on the platform (e.g., an accelerometer).

Sensors that measure external state include those that measure:

- the range from the robot to a point in its environment (e.g., sonar, radar or laser based range sensing);
- video imagery of the robot's surroundings (e.g., video camera);
- depth sensing of a portion of the video imagery (e.g., stereo camera).

Internal sensors such as odometers, gyroscopes, accelerometers and inclinometers can be used to compare how close the actual motion of a robot platform is in accordance with its motion commands. It is possible for example, that while carrying out a rotation motion on a wheeled platform the wheels may slip against the terrain. Evidence from the sensors that the platform did not move can be used to drive the motors to rotate an additional amount to compensate for the slip or to conclude that the platform is stuck. Similarly, driving a platform up a slope may result in a slower than expected forward velocity. If an inclinometer sensor can inform the robot that it is driving up a slope, then the robot can drive forward an additional amount to compensate for the effect of the slope.

External sensors can also be used to validate motion. We will deal with a specific example of this — localization using range information — in Chapter 6. They can furthermore be used to measure properties of the environment to select the best action to perform to achieve task objectives. Consider, for example, a mobile vehicle travelling along a road surface. Video information can be analyzed (using computer vision algorithms) to extract the direction of the road surface and steer the vehicle appropriately [3].

5.1. Transforming Sensor Readings

One of the most common varieties of exterioceptive sensor for a mobile robot is the *sonar range sensor*. Figure 5-1(a) and (b) shows the sonar configuration on the Pioneer model 3-AT robot. There are eight sonar sensors on the front (the small light-colored disks just under the top plate). Each sensor is a combined ultrasonic emitter-receiver. They operate as follows: An ultrasound ping is emitted and the time is measured to when the echo is returned. This configuration means that the distance to close objects — typically *6 cm* or less [2] — can't be measured since the emitter is still vibrating for an echo that close. The further the ping travels, the more it is attenuated. For the Pioneer 3-AT sonar, the longest reliable range is

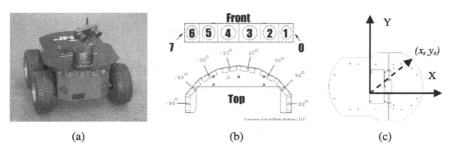

Figure 5-1: Pioneer 3-AT (a) Sonar configuration on 3-AT [5] (b) Local coordinate system (c).

about $4\,m$. The configuration of the front sonar array on the 3-AT is shown in Figure 5-1(b). Two sensors face sidewards while the other six are placed at fixed $20°$ increments around the front of the platform. There are eight similarly placed sensors on the back of the platform. As the robot moves, it can activate these sensors in sequence (typically 40 *ms* per sonar per array [5]) and build a $360°$ picture of its surroundings.

The 16 values produced by the front and back sonars are *range readings*, the distances from the reflecting object (in the best case, where the object is perpendicular to the sonar beam) to the sonar sensor. For some applications, this range format is sufficient. For example, if the robot is moving across an obstacle-strewn terrain, the range information can be used to stop the platform before it hits anything. However, for more involved interaction with the environment, it's useful to transform the range information into the location of a point in a Cartesian plane with origin at the center of the robot. Let (x_s, y_s) be the coordinates of the sonar sensor number s, where $s \in \{0, \ldots, 15\}$, in the local coordinate frame of the robot (Figure 5-1 (c)). Let the angle of the sonar axis with respect to the local X axis be ϕ_s. The location of the reflecting object (x, y), assumed to be on the center axis of the sonar beam (but see the next chapter for a more realistic treatment of this assumption), is given by

$$\begin{bmatrix} x \\ y \end{bmatrix} = \begin{bmatrix} x_s + r_s \cos \phi_s \\ y_s + r_s \sin \phi_s \end{bmatrix} \tag{5.1}$$

The angle ϕ_s is referred to as the *bearing* of the object and the distance r_s is referred to as the *range* of the object. Of course, the robot may be moving around, in which case the local coordinate frame will move with it.

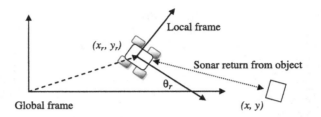

Figure 5-2: Translating sensor returns from local to global coordinates.

Thus, the same object will generate different local (x, y) coordinates depending on where the robot is.

It is useful therefore to consider a static global or world coordinate frame. Let us assume that the robot is initially positioned at the origin of the global frame and its local X and Y axes are aligned with the global axes. In that case, the object locations returned from evaluating eq. (5.1) for all $s \in \{0, \dots, 15\}$ are the same in local and global coordinates. Once the robot moves from the global origin to some pose (x_r, y_r, θ_r), the local and global coordinates are no longer the same. The situation is shown in Figure 5-2.

To translate the coordinates from the local to the global frame, the local frame needs to be rotated back through θ_r and then translated back (x_r, y_r). We can write this as follows:

$$\begin{bmatrix} x_g \\ y_g \end{bmatrix} = \begin{bmatrix} \cos\theta_r & -\sin\theta_r \\ \sin\theta_r & \cos\theta_r \end{bmatrix} \begin{bmatrix} x \\ y \end{bmatrix} + \begin{bmatrix} x_r \\ y_r \end{bmatrix} \tag{5.2}$$

where (x_g, y_g) are the coordinates of the object's location in the global frame. This can be written more succinctly using homogeneous coordinates as

$$\begin{bmatrix} x_g \\ y_g \\ 1 \end{bmatrix} = \begin{bmatrix} \cos\theta_r & -\sin\theta_r & x_r \\ \sin\theta_r & \cos\theta_r & y_r \\ 0 & 0 & 1 \end{bmatrix} \begin{bmatrix} x \\ y \\ 1 \end{bmatrix}$$

This operation needs to be completed for each sensor reading at each position of the robot. If there are n sensor readings, then we capture this with the following:

$$\begin{bmatrix} x_{0g} x_{1g} & x_{ng} \\ y_{0g} y_{1g} \cdots y_{ng} \\ 1 \quad 1 \quad\quad 1 \end{bmatrix} = T(x_{ri}, y_{ri}, \theta_{ri}) \begin{bmatrix} x_0 x_1 & x_n \\ y_0 y_1 \cdots y_n \\ 1 \quad 1 \quad\quad 1 \end{bmatrix} \tag{5.3}$$

where $T(x_{ri}, y_{ri}, \theta_{ri})$ for $i \in \{i, \ldots, m\}$ is the homogeneous transformation matrix for the ith robot position. As it happens, the calculations to transform the range readings from a laser ranger are basically the same as eqs. (5.1) and (5.3) except that n might be 180 rather than 16. If nm is large, then a parallel implementation of this transformation should be considered.

A serial implementation of this transformation will require n multiplications of a 3×3 matrix times a 3×1 matrix. To multiply a $p \times q$ matrix by a $q \times 1$ vector requires q multiplies and $(q - 1)$ additions for each of the p rows. If an addition operation takes a time a, and a multiplication operation takes a time b, then the matrix multiplication takes a time $p((q-1)a + qb)$. Thus, the serial computation time for the n multiplications of a 3×3 by 3×1 matrix is $3n(2a + 3b)$.

The design of a parallel implementation of this calculation should consider the ways in which the partitioning — how the data and calculation is spread over the cluster — can be implemented for this matrix expression: a 3×3 times a $3 \times n$ producing a $3 \times n$ in eq. (5.3).

5.1.1. *Partitioning: Single Robot Location*

Let us consider this problem for a single robot at a fixed location. The sensor readings, expressed in local coordinates, can be column distributed across all processors, resulting in $3n/p$ items per processor. The $T(x_r, y_r, \theta_r)$ matrix can be broadcast to all processors. Each processor then carries out n/p matrix-vector multiplications. Note: to broadcast T requires only broadcasting (x_r, y_r, θ_r); the 3×3 matrix can be generated from these values. Similarly, only $2n/p$ items need to be sent to each processor (since the $3rd$ item is always 1). This count remains the same if we choose to transmit the range and bearing information instead of the local Cartesian coordinates.

5.1.2. *Analysis*

The calculation time for this approach is the time to carry out n/p matrix vector multiplications. We don't need to do the last row of the matrix since it is always [0, 0, 1]; hence, $T_{cal} = 2(2a + 3b)(n/p)$ where a is the addition time and b the multiplication time as before. If we chose to transmit range and bearing then we would need to do one more multiply and one more add. We can assume that the *sin* and *cos* operations in eq. (5.1) can be accomplished with a table look-up operation since the bearing angles are known in advance (they are the sonar angles in Figure 5-1(b)).

The communication time will include:

- The broadcast of 3 values for the transformation matrix, taking time $\lambda \log p$ (because the message size is small and independent of n we will ignore the bandwidth related component of the communication cost).
- The scatter of $2n/p$ values for the local coordinates for the bearing and range values. In the latter case, we can assume that the sensor locations (x_s, y_s) will be constant and don't need to be transmitted. This will take $\lambda \log p + 2n(p-1)p\beta$.
- The gather of the $2n/p$ global coordinate values back to the root process, taking $\lambda \log p + 2n(p-1)/p\beta$.

The total communication time therefore is

$$T_{com} = 3\lambda \log p + 4n(p-1)/p\beta$$

and the total time is

$$T = \frac{2n}{p}(2a + 3b) + 3\lambda \log p + \frac{4n(p-1)}{p\beta}$$

We can explore extracting more parallelism from the matrix-vector loop: We can have each processor calculate only one row of the multiplication. The calculation time in that case is $T_{cal} = 2a + 3b$, a reduction by a factor of 2. However, we still have the broadcast of (x_r, y_r, θ_r), the scatter and then the gather of the $2n/p$ values, but now to 2 times as many processors, so the communication time increases.

Figure 5-3 shows the timing of the serial and column decomposition models for different numbers of sensor locations n. For small numbers of locations ($n = 50$), the parallelism incurs too much communication overhead, and the serial code is always faster. For a larger number of locations (*e.g.*, $n > 500$) the parallel code is faster. The shape of the column decomposition time for $n = 500$ shows a characteristic decline (calculation costs dominate the time for low values of p) followed by a slow rise (as communicate costs start to dominate for larger values of p).

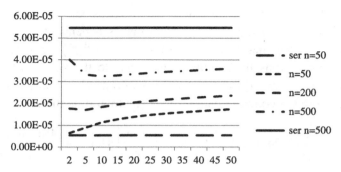

Figure 5-3: Parallel and serial sensor coordinate translation algorithm performance (time vs. number of processors) with one robot and one location showing serial versus parallel column decomposition times for different n ($\lambda = 3\,\mu s$, $\beta = 100\,Mbps$, $a = b = 7.3\,ns$).

For $n = 500$ the rise is very slow; the parallel time will eventually equal the serial time in this case when $p \cong 7500$ processors.

5.1.3. *Partitioning: Multiple Robot Locations*

So far we have considered the case of transforming sensor locations from local to global coordinates for a single location of a single robot. In reality, we may have sensor readings for many successive locations, and indeed we may have this data set for many robots.

Let us consider the case where there are multiple robots and each robot makes multiple moves. The transformation matrix for move i of robot j will be written $T(x_{ji}, y_{ji}, \theta_{ji})$ for $i \in \{1, \ldots, m\}$ and for $j \in \{1, \ldots, r\}$ and we consider eq. (5.3) modified appropriately. For simplicity we assume that for each position of each robot, there are always n sensor locations to be transformed.

The transformation operation for each position for each robot can be treated as an independent calculation. The serial calculation time is simply $T_{ser} = 2mrn(2a + 3b)$. Since all mr lists of n locations can be separately processed, we can have each processor calculate n/p transformations for every robot for every location. Thus the parallel calculation time $T_{cal} = 2mrn(2a + 3b)/p$ is the serial time divided by the number of processors.

However, we need to broadcast the mr transformation matrices to all processors, then we need to scatter the correct mrn/p elements to each processor, and finally, gather the elements back again from each processor. The communication time of the broadcast has, unlike the simple, single robot single location case, a term that is dependent on bandwidth (it requires 3 values per robot per position): $(\lambda + \frac{3mr}{\beta}) \log p$. The scatter and gather operations have similar time costs to the single robot single location case, but scaled by mr: $\lambda \log p + 2mrn(p-1)/p\beta$. These three terms compose the communication time. The total time therefore is:

$$T = \frac{2mrn}{p}(2a + 3b) + 3\lambda \log p + \frac{mr}{\beta} \left[3 \log p + \frac{4n(p-1)}{p} \right]$$

5.1.4. *Analysis*

Will any other partitioning improve on this result? Since there are mr independent calculations, we could decompose this problem into mr/p blocks. For each block, there are mr/p times n sensor locations to be translated. The calculation time is $T_{cal} = 2(2a + 3b)(mrn/p)$. The communication time includes a scatter of $3mr/p$ to each block to send the transformation matrices, followed by a scatter of $2mrn/p$ to each block to send the location data, followed by a gather of $2mrn/p$ from each block to reassemble the global location data. The total time in this case is slightly improved:

$$T = \frac{2mrn}{p}(2a + 3b) + 3\lambda \log p + \frac{mr(p-1)(3+4n)}{p\beta}$$

Figure 5-4(a) shows the predicted performance of this approach versus the serial approach for the case of a 20 robot team with each robot visiting 100 locations and making 200 measurements at each location. Figure 5-4(b) shows the graph of speedup versus number processors for the serial and parallel times of Figure 5-4(a). The diagonal $y = x$ is shown for reference. A speedup that is close to this diagonal is very attractive. In this case, the speedup is close to, but below, the diagonal. The drop off is not substantial. However, we can ask the question: is the drop off in speedup due to an algorithm with little scope for parallelism or is it due to the increase in communications overhead? The Karp-Flatt metric introduced in Chapter 3 can be used to tease out an answer.

Figure 5-4(c) shows the Karp-Flatt metric calculated for the speedup in Figure 5-4(b). Because the metric is continually increasing, the drop

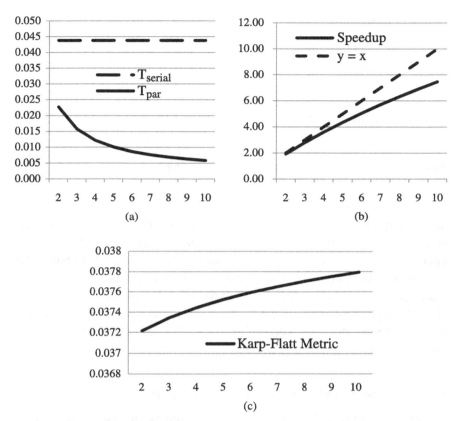

Figure 5-4: (a) Parallel and serial performance (time vs. num processors) with multiple robots and locations, (b) Speedup and (c) Karp-Flatt metric ($n = 200$, $m = 100$, $r = 20$, $\lambda = 3\ \mu s$, $\beta = 1 Gbps$, $a = b = 7.3\ ns$).

in speedup can be attributed to increasing parallel overhead and not to an algorithm with little scope for parallelism.

5.2. Drawing a Map from Sonar Data

Figure 5-5 below shows sonar data, in local and global coordinates, gathered on a short indoor trip by one robot ($r = 1$, $m = 12$, $n = 16$). The sonar configuration is that shown in Figure 5-1 for the Pioneer 3-AT robot.

The structure of the room becomes clear once the sensor return locations have been transformed from local coordinates (which are with respect to the moving robot platform) to global coordinates so that they are all

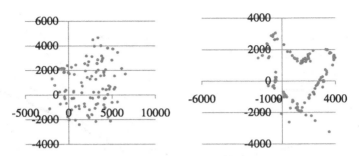

Figure 5-5:   Sonar data ($n = 16$) from an $m = 12$ location trip by one robot ($r = 1$). Left shows the data in local coordinates and right in global coordinates.

in the same reference frame. An important use of sensor returns such as this is the construction of a *map* of the area in which the robot is operating. The map can be used for many purposes (e.g., [2] chapter 8) including determination of this (or another) robot's location with respect to the map (localization), collaborative exploration [6] and mission planning [7]. Although the outline of the walls of the L-shaped area are clear to a human observer, the lines representing the walls need to be identified if this information is to be useful to a robot.

5.2.1. *Finding Straight Lines with the Hough Transform*

The *Hough transform* [14] is a method to automatically extract lines from a collection of n points. These points are represented by their Cartesian (x, y) coordinates. Each line must contain at least two points, in which case the two points can be used to calculate the slope m and offset c for the line equation: $y = mx + c$. The Hough transform takes every pair of points and calculates the pair (m, c) for those points. A count is kept for each point in the new space (m, c); those points with higher counts represent more pairs of (x, y) points. Because each point is compared with each other point, this method has $O(n^2)$ complexity.

Hough transforms are commonly used in image processing, and parallel formulations of them exist, e.g., [4,13]. We will use polar form for the line equation:

$$x \cos \propto + y \sin \propto - P = 0 \qquad (5.4)$$

In this case, P is the distance from the line to the origin and α is the angle that a perpendicular to the line through the origin makes with the x axis. It's important to consider both the range and the quantization of the (α, P) transform space. Since our sonar sensors are only accurate to $4\,m$ at most from the robot, we can establish the range $\{-4000, \ldots, 4000\}$ in *mm* for P. The angle α can extend from 0 through 2π, of course. Note that since P can be both negative and positive, we will hit every line with two angles, π radians apart. For this reason, we will limit α to the range $\{-\pi/2, \ldots, \pi/2\}$. We need to pick a quantization factor for both:

$$n_P = 8000/q_P$$
$$n_\alpha = \pi/q_\alpha$$

These establish the size of the 2D *accumulator* array that stores the evidence counts for each point in the space. Let's assume that the points are stored in an array points[n]. Exploiting symmetry, we process each of the data points points[i] with each of the other data point points[j] for $j \in \{i, \ldots, n - 1\}$ giving a total $n(n - 1)/2$ operations. If we use a 1D array for *accumulator* then we can access each point (i_P, i_α) in our quantized 2D Hough space as:

$$accumulator[i_P][i_\alpha] = accumulator\ 1D[n_P * i_\alpha + i_P].$$

Using a 1D array has the advantage of making it easier to sort the count values once they have been calculated. Give any two points $(x1, y1)$ and $(x2, y2)$ we can calculate the values for P and α as follows:

$$\propto = a\tan 2(y2 - y1, x2 - x1)$$
$$P = x1\cos\propto + y1\sin\propto$$

5.2.2. *Partitioning*

It would be convenient to do partitioning by dividing the *accumulator* array by row block, or by column block, over the cluster. However, we cannot tell in advance which of the n points will contribute to what parts of the $n_\alpha \times n_P$ array, and hence we could not easily direct data points to the correct processor. Furthermore, it is unlikely that all parts of the Hough space will be equally covered by the data, so this partitioning could produce unbalanced processor loads. Instead, we have to settle for partitioning n/p

points to each processor. If however, each processor only tests its n/p points against themselves, then many potential lines may be missed. Each processor needs to test n/p points against all n points to ensure no lines are missed.

5.2.3. *Program Design*

The parallel algorithm begins by broadcasting all the points to all processors. The following is the function prototype for MPI_Bcast:

```
int MPI_Bcast (
                void            *buffer,
                int             count,
                MPI_Datatype    datatype,
                int             root,
                MPI_Comm        comm    ).
```

The parameters of MPI_Bcast are similar to the parameters for the scatter and gather operations that we have already covered. All count values in the buffer on the root processor are transmitted to the local copy of buffer on each processor. We broadcast the n data points with:

```
MPI_Bcast ( points, n, MPI_FLOAT, 0, MPI_COMM_WORLD );
```

Each processor will process its portion of the points array using the BLOCK_LOW and BLOCK_HIGH definitions from the previous chapter. This code is shown in Figure 5-6.

The accumulator array will need to be initialized by each processor. The code in Figure 5-6 will fill in parts of the array on each processor.

```
for (i = BLOCK_LOW(id,p,n); i <= BLOCK_HIGH(id,p,n); i++ )
    for (j = i+1; j < n; j++ ) { /* get polar form of line */
        alpha = atan2( points[j].y-points[i].y,
                       points[j].x-points[i].x);
        P  = points[i].x*cos(alpha)+points[i].y*sin(alpha);
        /* then get Hough space index */
        i_alpha = (M_PI + alpha) / q_alpha;
        i_P     = (4000 + P) / q_P;
        /* increment accumulator for this Hough space point */
        accumulator1D [ n_P * i_alpha+ i_P ]++;
    }
```

Figure 5-6: Body of parallel Hough transform.

The accumulator array then needs to be reduced by summation on the root processor so that the evidence counts for the portions of the Hough space on each processor can be added for a final total.

```
MPI_Reduce( &accumulator1D, &g_accumulator1D, n_alpha*n_P,
            MPI_INT, MPI_SUM, 0, MPI_COMM_WORLD);
```

The root needs to sort the accumulator values to find those lines that have the most support (i.e., the largest accumulator values). Sorting the 1D array used to this point is problematic, since the index information (the Hough space coordinates) is lost when the array is sorted. Instead we can modify our 1D accumulator to consist of not just the count value, but also the Hough space coordinates:

```
struct {
        int count; /* number of lines with alpha and P */
        float alpha; /* hough space coordinates of array cell */
        float P;
} accumulator1D[ n_alpha * n_P ];
```

Now when the array is sorted, the Hough space coordinates 'stick' with their original count. The sorting can be done on the root processor with a time complexity $n_P n_\alpha \log n_P n_\alpha$. Sorting is itself a topic of interest for parallelism — see Quinn [8] Chapter 14 for several parallel sorting algorithms. Figure 5-7 shows the output from the algorithm for the sonar dataset in Figure 5-5.

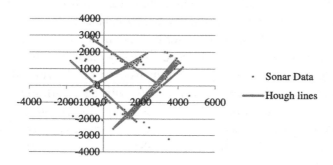

Figure 5-7: Top 30 Hough lines for the dataset in Figure 5-5.

5.2.4. *Analysis*

Let χ be the cost of executing the body of the loop in Figure 5-6. The serial algorithm does $n^2/2$ operations, in time $n^2\chi/2$. However, the processors allocated to low blocks of the array of points have more work to do than processors allocated to later blocks, and the processor allocated to the first block has the biggest load of all. Hence we are obliged to consider the worst case condition for the parallel calculation time T_{cal}.

The loop in Figure 5-6 will, in the *worst case*, do $\frac{1}{2} \times \frac{n}{p} \times \frac{n}{p} = n^2/2p^2$ operations on data in its allocated group of n/p points. Subsequently, it will do $(p-1)$ times n^2/p in processing its group of n/p points with the remaining $(p-1)n/p$ points. This time is much worse than a linear speedup $n^2\chi/2p$. Indeed as p increases, the parallel loop time does approache $n^2\chi/2p$. This is because a processor can only exploit the symmetry for the n/p points in its block, so T_{cal} approaches $n^2\chi/2p$ for large p since the vast majority of points will not be in the processor's block.

The communication costs include a broadcast of n points costing $(\lambda + 2n/\beta)\log p$ (assuming an x and y ordinate for each point) and a reduce cost of $n_P n_\alpha$ times three values costing $(\lambda + 3n_P n_\alpha/\beta)\log p$. The reduction time is essentially fixed by the selection of the quantization of Hough space. The total time is therefore:

$$T = \frac{n^2}{2p}\left[\frac{1}{p} + \frac{2(p-1)}{p}\right]\chi + \left[2\lambda + \frac{2n + 3n_P n_\alpha}{\beta}\right]\log p$$

5.2.5. *Load Balanced Hough Calculation*

The calculation time is the dominant term in this expression, so it's natural to wonder if there is a way to reduce this by balancing the processor load. The problem is that the first processor carries out an $n^2/2p^2$ calculation on 'its' assigned block — but it still needs to carry out n/p times $(p-1)n/p$ more calculations, processing points in its n/p block with all n points. Figure 5-8 shows an example of n points with data partitioned across 5 processors. Each processor is allocated its group of n/p points — these are the shaded diagonal blocks in the Figure. Each processor does an $n^2/2p^2$ calculation in 'its' diagonal block and then a kn^2/p^2 calculation for the

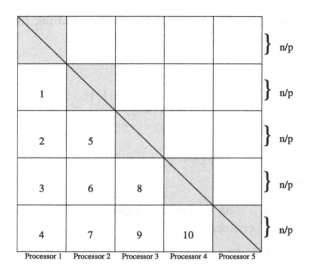

Figure 5-8: Example Hough space data partitioning for 5 processors.

remaining points in that column, where $k \in \{0, \ldots, 4\}$ for this case of 5 processors.

We can *balance the load* on each processor by reassigning the blocks numbered 1 through 10 in Figure 5-8 to other processors. Each processor can do its assigned $n^2/2p^2$ block, and then some balanced number (in this case 2) of n^2/p^2 blocks. If there are p processors then the length of the diagonal will be p and there will be $p(p-1)/2$ blocks below the diagonal, numbered as shown for $p = 5$ in Figure 5-8. We want to allocate $p(p-1)/2$ blocks to p processors, therefore each processor gets $(p-1)/2$ blocks.

By the same reasoning used in the block decomposition approach of Chapter 3, we can say that in the case where $(p-1)/2$ has a fractional component, the unequal block allocations should be distributed across p rather than being all at the low, or all at the high numbered processors. The Block Decomposition Macros can be used to carry this out.

Given a block number in the scheme shown in Figure 5-8, we need to determine the index ranges in $\{0, \ldots, n-1\}$ for the points in this block. The following code will evaluate the two starting indices index1 and index2 for a block:

```
col = 0;
for ( i = 0; i < p − 1; i++ )
        if ( ( block−col ) ≤ ( p − i ) ) {
                index1 = i ;
                index2 = i + ( block + col ) − 1;
        }
        else col += p − i ;
```

5.2.6. *Analysis*

Assuming $(p - 1)/2$ is an integer for simplicity, then the calculation time for each processor becomes:

$$T_{cal} = \left[\frac{n^2}{2p^2} + \frac{(p-1)n^2}{2p^2} \right] \chi = \frac{n^2}{2p} \left[\frac{1}{p} + \frac{(p-1)}{p} \right] \chi = \frac{n^2}{2p} \chi$$

The communication time stays the same. This balanced algorithm performs far better than the unbalanced, as shown in the graph of Figure 5-9: The balanced execution time is a major improvement on the serial time, whereas the unbalanced time is a lot worse than the serial time.

5.3. Aligning Laser Scan Measurements

The SICK™ LMS 200 laser rangefinder (see Figure 5-10(a)) uses a time-of-flight (LIDAR, light radar) approach to measure distances. We will

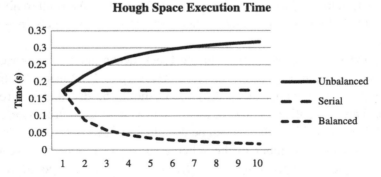

Figure 5-9: Graph of serial versus unbalanced and balanced parallel performance. ($n = 1000$, $\chi = 350\,ns$, $n_P = n_\alpha = 100$, $\lambda = 3\,\mu s$, $\beta = 1\,Gbps$, $a = b = 7.3\,ns$).

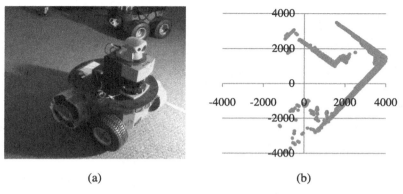

(a)   (b)

Figure 5-10: (a) Pioneer 3-AT robot with SICK LMS 200 laser rangefinder; (b) Laser rangefinder data.

assume a configuration of the LMS 200 in which it produces a series of 180 measurements per scan. Each measurement is the range from the center of rotation of the laser scanning mirror in the LMS to the point in the environment reflecting the laser back. The measurements are at one degree increments. Figure 5-10(b) shows a dead-reckoning combination of 12 movements each with 180 sensor measurements by one robot. (In fact, this is the same location, robot and path that yielded the sonar measurements in Figure 5-5.) The SICK LMS 200 has a range of between $10\,m$ and $80\,m$ depending on the reflectivity of the surfaces being measured. The angular resolution can be configured at $1°$, $0.5°$ or $0.25°$ and the range resolution is $1\,mm$. Given the accuracy of the sensor, the dead-reckoning combined readings in Figure 5-10(b) seem surprising: there appears to be quite a spread of range values along the walls as shown in the close-up in Figure 5-11(a). This is not an issue with the sensor, but with the dead-reckoning approach. That approach uses the odometry information from the robot to transform the sensor data into global coordinates. The robot odometry, however, is error prone [9]; especially so when the robot rotates. Certainly the odometry provides a good starting point for aligning the scan data from each of the 12 scans, but additional processing is necessary to refine the match.

There are several approaches to laser scan matching. One of the most popular is the Iterative Closest Point (ICP) approach of [10]: The closest points in two successive scans are matched and the robot position and

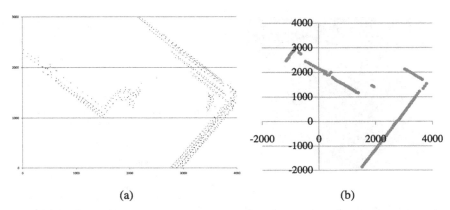

(a)   (b)

Figure 5-11: (a) Closeup of Fig. 5-10(b) showing misregistration; and (b) after Polar-Scan Matching.

orientation calculated from this iteratively until a termination criterion is met. Nuchter [11] describes a parallel version of ICP designed for a multiprocessor shared memory system which showed a 48.2% improvement over the serial algorithm for 3D laser datasets. We will look at a parallel version of an approach to matching 2D laser scans proposed by Diosi and Kleeman [12] called *polar scan matching (PSM)*. Figure 5-11(b) shows the result of using polar scan matching on the 12 scans from Figure 5-10(b).

Polar scan matching takes advantage of the fact that the laser range data is gathered in polar form (that is, bearing and range). Whereas ICP tries to match closest points (and hence matches every point with every other point), PSM matches points with the same bearing.

5.3.1. *Polar Scan Matching*

We will represent a single laser scan in the following way:

$$C = (x_c, y_c, \theta_c, \{(r_i, \theta_i)|i = 1, \ldots, n\})$$

where the center and orientation of the laser ranger is given by the pose (x_c, y_c, θ_c) and where there are n range readings given by the set of range and angle pairs $\{(r_i, \phi_i)|i = 1, \ldots, n\}$. For our SICK laser scanner problem, $i = 1, \ldots, 180$ and $\phi_i = 1°$. The *scan matching problem* is defined as matching a scan C against a reference scan C_r so that both scans are aligned. (Figure 5-11(a) shows scan misalignment.) There are three main steps in Polar Scan Matching, and these three steps are iterated until the alignment change is sufficiently small.

Step 1. The first step is to determine what C would look like if it were viewed from the location of C_r. This is called *scan projection*. Interpolation is necessary to match the projected scan endpoints from C with readings taken along the bearings from C_r. Once C has been projected to the location of C_r — lets call this $\tilde{C}$ — the position and the orientation of the scan with respect to C_r can be calculated.

Step 2. The position correction $(\Delta x_c, \Delta y_c)$ is calculated by minimizing the sum of the weighted range residuals (weighted least-squares):

$$\begin{bmatrix} \Delta x_c \\ \Delta y_c \end{bmatrix} = (\boldsymbol{H}^T \boldsymbol{W} \boldsymbol{H})^{-1} \boldsymbol{H}^T \boldsymbol{W} (\tilde{\boldsymbol{r}}_c - \boldsymbol{r}_r) \tag{5.5}$$

Here $(\tilde{\boldsymbol{r}}_c - \boldsymbol{r}_r)$ is the vector of differences between reference and scan projected ranges. The matrix $\boldsymbol{H}$ specifies the polar relationship between the position $(\Delta x_c, \Delta y_c)$ and each bearing and range:

$$\boldsymbol{H} = \begin{bmatrix} \dfrac{\partial \tilde{r}_{c1}}{\partial x_c} & \dfrac{\partial \tilde{r}_{c1}}{\partial y_c} \\ \dfrac{\partial \tilde{r}_{c2}}{\partial x_c} & \dfrac{\partial \tilde{r}_{c2}}{\partial y_c} \\ \cdots & \cdots \end{bmatrix} = \begin{bmatrix} \cos(\varphi_{r1}) & \sin(\varphi_{r1}) \\ \cos(\varphi_{r2}) & \sin(\varphi_{r2}) \\ \cdots & \cdots \end{bmatrix}$$

$\boldsymbol{W}$ is a matrix of weights used to reduce the effect of bad matches, giving a small weight to large residuals.

Step 3. The orientation correction is obtained by rotating the projected scan $\tilde{C}$ in small angular increments and calculating the range residual. The scan C rotated by α is given by

$$\begin{aligned} C/\alpha &= (x_c, y_c, \theta_c + \alpha, \{(r_i, \phi_i) | i = 1, \ldots, n\}) \\ &= (x_c, y_c, \theta_c, \{(r_i, \phi_i + \alpha) | i = 1, \ldots, n\}) \\ &= (x_c, y_c, \theta_c, \{(r_{(i+\alpha)}, \phi_i) | i = 1, \ldots, n\}) \end{aligned}$$

The result of adding α to the ranges in C/α is to reorder the ranges. For example, adding $\alpha = 1°$ would mean that range r_1 now becomes r_2 and so

forth. To calculate the range residuals we still just subtract ranges, but only
those with the same index:

$$e(\propto) = \sum_{i=\propto}^{n}(r_r - r_c / \propto) \qquad (5.6)$$

For a negative angle, the summation indices are from $n - \alpha$ to 1. The
error function e is minimized over an angular displacement window to
obtain the orientation correction.

Steps 1 and 2 consist of one or more calculations that need to be carried
out for each (r_i, ϕ_i) in the scan, independent for the most part of each
other point in the scan. In Step 1 there is additionally the interpolation
calculations, and in Step 2 there is additionally a summation reduction in
the weighted least squares calculation. However, these calculations can be
distributed across cluster processors in a manner similar to the approaches
for the first algorithm in this chapter. Step 3 is different: if the rotation
window is of width m degrees, then approximately mn calculations are
necessary to minimize the error function in eq. (5.6).

5.3.2. *Partitioning and Analysis*

As always, we explore the implications of distributing the data and the
calculations in different ways over the cluster. The calculations can be
partitioned in two main ways:

(1) Each processor can carry out a subset m/p of the window calculations
    for all n bearings. Both C and C_r need to be broadcast to all
    processors, a communication of $2(2n + 3)$ values. Each processor
    will carry out the nm/p range residual differences and summations
    to calculate the m/p values of the error function. In fact this is an over
    estimate, but we will let it stand for now. The minimum of the m/p
    errors then needs to be found. This will take a time proportional to
    m/p. Finally the p minima need to be reduced to a global minimum —
    a reduction operation that takes logarithmic time.
(2) Each processor can carry out all m window calculations for n/p
    bearings. This requires that the $2n/p + 3$ values from the reference scan
    be transmitted to each processor, but that $[2n/p + 2m] + 3$ values of

the scan to be matched are transmitted in order to allow the window to 'slide' down m range and bearing values. The calculations are mn/p range residuals, each of which is just a portion of the m values of the error function; these m values need to be reduced by summation, followed by a linear search for the global maximum.

Comparing these two approaches, we can see that the calculation time for the first approach is slightly better, since only a portion of the error function needs to be searched in a linear fashion for the minimum. The communication time for each will include a $3\lambda \log p$ term since both carry out three group communication operations: two broadcasts and a reduce, or two scatters and a reduce, respectively.

Simplifying slightly, for the first approach, the two broadcasts take $2[\lambda + 2n/\beta] \log p$ and the final reduction $\lambda \log p$. Simplifying again, the second approach takes $2\lambda \log p + 4n(p-1)/p\beta$ for the two scatters, but also an additional $2m(p-1)/p\beta$ for the 'window' to slide down the scan being registered on the second scatter, and $[\lambda + m/\beta] \log p$ for the final reduce operation.

Dropping the $3\lambda \log p$ from both times, the first approach has the larger communication time iff:

$$\frac{4n}{\beta} \log p \; > \; \frac{(p-1)(4n+2m)}{p\beta} + \frac{m}{\beta} \log p$$
$$\implies \frac{4n-m}{4n+2m} \log p > \frac{p-1}{p}$$

Given that $0 \le m \le n$ (and typically $m << n$) and that p must be an integer greater than 1, this condition holds for all values of p. Hence, the second approach has the better (smaller or equal) communication time.

5.3.3. *Program Design*

Implementing the second approach means we have to scatter overlapped blocks of the scan to be registered. Up until now we have only scattered *non-overlapping* blocks of data. Let us assume that the range and bearing information for the reference scan is stored in range_r, phi_r. The range and bearing for the scan to be registered is stored in range and phi. We will again make the assumption that n is a multiple of p, and block scatter the reference scan with the following code:

```
MPI_Scatter( &range_r[0], n/p, MPI_FLOAT,
                 &range_rl[0], n/p,MPI_FLOAT,
                 0,MPI_COMM_WORLD );
MPI_Scatter( &phi_r[0], n/p, MPI_FLOAT,
                 &phi_rl[0], n/p,MPI_FLOAT,
                 0,MPI_COMM_WORLD );
```

We need to introduce a new MPI operation to scatter the scan to be registered in overlapping blocks. The following is the function prototype for **MPI_Scatterv**:

```
int MPI_Scatterv (    void            *sendbuf,
                      int             *sendcnts,
                      int             *displs,
                      MPI_Datatype    sendtype,
                      void            *recvbuf,
                      int             recvcnt,
                      MPI_Datatype    recvtype,
                      int             root,
                      MPI_Comm        comm    ).
```

As with **MPI_Scatter**, the **sendbuf** parameter is the address of the array (buffer) being scattered and **recvbuf** is the address of the array (buffer) into which the scatter elements are to be placed. However, **MPI_Scatterv** allows for varying amounts of data to be scattered to each processor. The **displs** and **sendcnts** parameters are arrays containing the index and size of the block to be sent to each processor. Note additionally that this operation offers us a way out of the assumption that we have been making in this and previous chapters: that n is a multiple of p. The following code segment assigns the block start address and size correctly according to Quinn's block decomposition scheme from Chapter 4 for use in **MPI_Scatterv**:

```
for (i= 0; i < p; i++) {
        displ[i ]    = BLOCK_LOW( id, p, n );
        dendcnt[i ] = BLOCK_SIZE( id, p, n );
}
```

We can use this same code segment to send overlapping blocks of the array, we just need to make the block lengths *larger* than BLOCK_SIZE(id,p,n) by the size of the search window:

$$\text{dendcnt[i] = BLOCK\_SIZE(id, p, n) + m.}$$

The angular search window needs to go $m/2$ in a positive direction and then $m/2$ in a negative direction, not m in just the positive (or just the negative direction). Hence, while we need to extend the length of the section sent by m as discussed, we also need to move the start address *back* by $m/2$:

```
for ( i = 0; i < p; i++ ) {
        displ[ i ]   = max( 0, BLOCK_LOW(id, p, n) – m/2);
        dendcnt[ i ] = min( BLOCK_SIZE(id, p, n) + m, n – displ[ i ]);
}
```

To be sure we don't send a negative address, we use the **max** function of our address calculation with 0. And to be sure we don't attempt to send more elements than are left in the array, we use a min function with the amount of elements left in the array. We can now scatter the range (and similarly for bearing) information with:

```
MPI_Scatterv( &range[0], displ, dendcnt, MPI_FLOAT,
              &range_l[0], dendcnt[rank], MPI_FLOAT,
              0, MPI_COMM_WORLD );
```

Once the data has been scattered to the local buffers, each processor can calculate its portion of the error function. For the middle part of the range arrays, range_rl[0] will correspond with range_l[$m/2$] and the residual calculations are straightforward. However, for the start and end of the array, the full BLOCK_SIZE(id, p, n) may not be available for calculation of the residuals, so processors that are allocated those parts of the array need to ensure that they check the scan indexes for these 'edge effects' as indicated in the code section below.

```
if ( BLOCK_LOW(id, p, n) < m/2 )
        offset = m/2 - BLOCK_LOW(id, p, n);
else
        offset = m/2;
for ( w = -m/2; w < m/2; w++ ) { /* w is the angular displacement */
        residual = 0.0;
        c = 0; /* will count residual calculations */
        for ( i = 0; i < BLOCK_SIZE(id, p, n); i++ ) {
                index = i + w + offset; /* index into scan to be reg. */
                if (index>0 && index < dendcnt[rank] ) { /* edge*/
                        residual += range_rl[ i ] -  range_l[ index ];
                        c++;
                }
        }
        error_l[ w ] = residual / c;
}
```

In the case that the full **BLOCK_SIZE(id, p, n)** is not available, the number of valid residual calculations is counted (in c) and used to normalize the error value. Finally, the error array needs to be reduced by summation over all processors and the minimum error value found:

```
MPI_Reduce( &error_l, &error, m, MPI_FLOAT, MPI_SUM, 0,
            MPI_COMM_WORLD);
If ( rank == 0 ) {   /* root processor only*/
                min_error = error[ 0 ]; min_index = 0;
                for ( i = 1; i < m; i++ )
                        if ( error[ i ] < min_error ) {
                                min_error = error[ i ];
                                min_index = i;
                        }
                new_theta = old_theta + radians( min_index - m/2 )
}
```

The index of the minimum error gives the change in orientation angle for the best alignment.

5.4. Summary

In this chapter, we have looked at some of the problems involved in processing sensor information. We covered three problems:

(1) Translating the location of sensor readings from local to global coordinates for a single robot and for multiple measurements from a team of robots.
(2) Extracting straight lines from sonar (but it could as easily have been laser [15]) data using a Hough transform.
(3) Scan matching laser scans using the Polar Scan Matching method.

In the Hough transform case, we considered how to design the data partitioning so that the load on the processors in the cluster is balanced. This can have a major effect on performance — in the Hough transform case, the unbalanced parallel algorithm is *always* worse than a serial implementation.

We used the Polar Matching method to address the laser scan alignment problem. We introduced the MPI_Scatterv command, an MPI command which allows us to scatter variable sized, and possibly overlapping, data blocks to processors. In particular, this command allows us to implement Quinn's data decomposition macros, first discussed in Chapter 4, relaxing the assumption that the amount of data to be scattered is always a multiple of the number of processors.

Another approach to scan alignment that maps well to a cluster implementation places 'springs' or 'virtual forces' between scans. The scans are iteratively moved under the influence of the forces so that they are 'pulled' into alignment with one another. One example is the Force Field Simulation method of Lakaemper *et al.* [16]. These approaches are modeled on the *n-body* problem in physics and can be addressed therefore in a manner similar to parallel *n-body* simulations (see [8], Chapter 6).

References

1. Arbib, M.A., *The Metaphorical Brain 2*, Wiley & Sons 1989.
2. Dudek, G. and Jenkin, M., *Computational Principles of Mobile Robotics*, Cambridge Press 2000.
3. Pomerleau, D., Neural Networks for Intelligent Vehicles, *Proceedings of the Intelligent Vehicles Conference*, Tokyo, Japan 1993.

4. Braunl, T., Feyrer, S., Rapf, W., and Reinhardt, M., *Parallel Image Processing*, Springer-Verlag, 2001.
5. Mobilerobots Inc. Pioneer 3 & Pioneer 2 H8-Series Operations Manual, 2003.
6. Fox, D., Ko, J., Konolige, K., Limketkai, B., Schulz, D., and Stewart, B., Distributed Multi-Robot Exploration and Mapping, *Proc. of the IEEE: Special Issue, Multi-Robot Systems V94 N7, 2006.*
7. Brummit, B. and Stentz, A., Dynamic Mission Planning for Multiple Mobile Robots, *Proc. IEEE Int. Conf. on Robotics and Automation*, Minneapolis MN, 1996.
8. Quinn, M.J., *Parallel Programming in C with MPI and OpenMP*, McGraw-Hill 2004.
9. Borenstein, J., Experimental Results from Internal Odometry Error Correction with the OmniMate Mobile Robot, *IEEE Trans. on Robotics and Automation, Vol. 14, No. 6*, December 1998, pp. 963–969.
10. Besl, P. and McKay, N., A Method for Registration of 3–D Shapes. *IEEE Trans. on PAMI*, 14(2):239–256, February 1992.
11. Nuchter, A., Parallelization of Scan Matching for Robotic 3D Mapping, *Proc. 3rd European Conf. on Mobile Robots*, 2007.
12. Diosi, A. and Kleeman, L., Laser Scan Matching in Polar Coordinates. *Proc. Int. Robots and Systems (IROS)*, Edmonton, Canada 2005.
13. Wilkinson, B. and Alen, M., *Parallel Programming*, 2nd Ed., Prentice-Hall 2005.
14. Hough, P.V.C. Methods and means for recognizing complex patterns, *US Patent #3,069,654.* 1962.
15. Sack, D. and Burgard, W., A comparison of methods for line extraction from range data, 5th *IFAC Symposium on Intelligent Autonomous Vehicles*, Lisbon, Portugal, 2004.
16. Lakaemper, R., Adluru, N., Latecki, L.J., and Madhavan, R., Multi Robot Mapping using Force Field Simulation. *Journal of Field Robotics*, 24(8/9), 2007, pp. 747–762.

Chapter 6

Mapping and Localization

It is possible for a mobile robot to navigate with little knowledge of its environment.[a] However, having a map of the area to be explored is of tremendous use, since potential obstacles can be seen in advance, and more efficient motions (in terms of time or energy expended) can be carried out. Anyone who has faced the daunting task of running multiple errands in a large shopping mall can appreciate the value of a good mall map in saving time and sore legs. Even if the map is only partial, it still provides a boundary between what is known and what is not yet known, and it can be used to steer additional map making. However, even the best map is of no value unless the 'you are here' symbol can be found. Given a map, the first action a robot needs to perform therefore is to identify its own location on the map. The process of making a map from robot sensor data is called *mapping* and the process of determining where the robot is with respect to the features on the map is called *localization*. We will look at both of these problems in this chapter.

Two common categories of robot map are:

(1) **A Metric map:** A collection of absolute spatial measurements between landmarks identified by sensory feature measurements. The robot can use its sensors to identify what landmarks are within its sensor range, and by comparing these to the information in the map, can determine where it is spatially on the map (*localization*). When the robot has localized itself, it can then plan its motions accurately using

---

[a]The *Bug* algorithms discussed in [3] are an example of this.

the spatial measurements from the map. A metric map is similar to a place map as seen on Google™ maps.

(2) **A Topological map:** A graph-like collection of places, each identified by a collection of sensory feature measurements (the place *signature*), and linked together by routes without any absolute spatial measurements. A robot can determine at which place it is by matching its sensor readings with the place signatures. The route information allows the robot to transition from one place to another. However, motor skills or behaviors (e.g., obstacle avoidance, lane following) may be necessary to successfully navigate the transition from one place to another. A topological map is more similar to a subway map than to a Google™ map.

Each approach has its own strengths: The metric map allows for accurate motion planning, but at the cost of representing a lot of detail about the environment (not all of which may be relevant for any particular task). The overhead of representing this detail may limit the amount of space that can be represented by a metric map. The topological map omits metric detail and hence is less sensitive to the change in geometric details in places (e.g., a chair moved from one spot to another in a room) and is also suitable for representing larger spaces. To integrate the advantages of each approach, Dudek [5] proposes a categorization of map data representations into five hierarchical levels:

(1) **Sensorial:** The map is composed of raw sensor data.
(2) **Geometric:** The sensor data is used to infer the three-dimensional objects that compose the map.
(3) **Local relational:** The objects are annotated with the functional, structural or semantic relations to their neighbors.
(4) **Topological:** Objects are grouped into large scale locations that are connected together if it is possible to travel from one location to the other.
(5) **Semantic:** Functional labels are associated with the map elements.

One of the earliest forms of sensorial map developed was the *spatial occupancy grid (or map)*. The occupancy grid represents space as a collection of discrete spatial cells. Figure 6-1 shows an example occupancy grid map of a room with a doorway and a short entrance hallway as well as

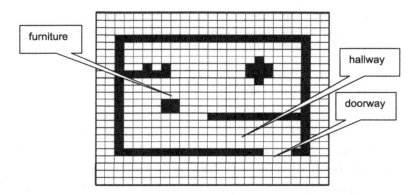

Figure 6-1: Occupancy grid representation of space.

several pieces of furniture. The cells colored in black represent occupied space on the map and those uncolored represent free space.

Each grid cell in Figure 6-1 represents an area of space centered at some two-dimensional floor location (x, y) with respect to a world coordinate frame. In the earliest forms of occupancy grid, the cell on the map corresponding to the location (x, y) was binary: either *occupied* if there was an object occupying any portion of that space, or *unoccupied* if the space was free. In later versions of the occupancy grid (e.g., [6]), the value of the cell at (x, y) contained the probability of occupancy.

6.1. Constructing a Spatial Occupancy Map

The spatial occupancy map can be constructed from any sensor input that provides range or depth measurements. It is very common to construct an occupancy map from sonar measurements for example. In the previous chapter we have discussed the sonar configuration of the Pioneer 3-AT robot (Figure 5-1) and we discussed approaches to mapping sonar sensor range and bearing information to a global coordinate frame. We assume in eq. (5.1) that when a sonar s mounted on the robot returned a sensor reading with range r_s and bearing ϕ_s that the reflecting object was located a distance r_s away along the bearing ϕ_s. While this is a reasonable assumption for laser range-finding, it is unrealistic for sonar. A sonar beam will spread outwards from the sensor in a cone. For the SensComp 600^{TM} sonars mounted on the Pioneer 3-AT, the angle of the cone is 15°. A strong reflection anywhere in this cone may be responsible for our range reading. Moravec & Elfes [8] introduced a sonar probability model to capture this uncertainty.

6.1.1. *Probabilistic Sonar Model*

Figure 6-2(a) shows a cross section of a sonar beam overlaid on an occupancy grid. The beam can be divided into four areas labeled R1 through R4. R1 is the area between the robot and the sonar reading; R2 is the area of the beam about the same range as the reading (shown containing a black occupancy grid cell); R3 is the area of the beam beyond this; and, R4 is all the area outside the beam. We cannot say anything about R4 from this sonar measurement. We restrict our attention to R1 through R3. R1 is probably empty, though that probability decreases as we get closer to the range reading. R2 is probably occupied but as we get further (in range and bearing) from the range reading that probability decreases. R3 is occluded by the range reading, so we cannot say anything about it.

Figure 6-2(b) shows a probabilistic model that fits these requirements. There is a valley in R1 shrinking in size as it gets closer to the range reading in R2, showing a lower probability of occupancy (higher probability of being empty). There is a peak in R2 at the range reading and the probability of being occupied falls off towards the boundaries of the R2 region.

$$p_{R2}(occupied) = \frac{\frac{R-r}{R} + \frac{\beta - \alpha}{\beta}}{2} \tag{6.1}$$

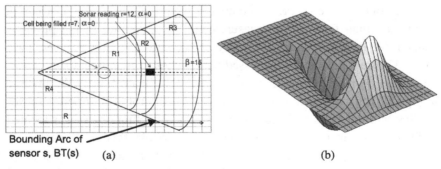

Figure 6-2: (a) Probabilistic sonar model superimposed on occupancy grid (b) Combined region R1 & R2 sonar probability model (Moravec & Elfes [8]).

Equation (6.1) approximates the probability of occupancy in R2 for a cell that is at distance r from the sonar and at angle α to the axis of the sonar. This same expression can be used to generate $p_{R1}(empty)$ the probability of being empty for a cell at (r, α) in region R1.

6.1.2. *Bayesian Filtering*

Any particular cell in the occupancy grid may repeatedly fall under the beam of a given sonar sensor as the robot moves, and of course, it may also fall under the beams of other sonar sensors. Thus, we need a way to recursively update the probability of occupancy to reflect each such additional observation from the probabilistic model in Figure 6-2(b). Considering a single occupancy grid cell: let us denote by H the hypothesis that the cell is occupied and by $\neg H$ the hypothesis that is empty. Let $p(H)$ be the probability that the cell is occupied and $p(\neg H)$ the probability that it is empty. We insist that $p(H) = 1.0 - p(\neg H)$; that is, a cell is either empty or occupied.

The conditional probability $p(H|s)$ is the probability that the cell is occupied given the sensor reading s. In fact we may have a series of readings $s_1, \ldots, s_n$ from the same sonar or from other sonar sensors that have covered this cell, and we want to calculate the probability of occupancy based on all those readings: $p(H|s_n, s_{n-1}, \ldots, s_1)$. Bayes rule allows us to state this as follows:

$$p(H|s_n, s_{n-1}, \ldots, s_1)$$
$$= \frac{p(s_n, s_{n-1}, \ldots, s_1|H)p(H)}{p(s_n, s_{n-1}, \ldots, s_1|H)p(H) + p(s_n, s_{n-1}, \ldots, s_1|\neg H)p(\neg H)}$$

The bottom line above is the total likelihood. If we make an assumption that the sequence of sensor readings is independent[b] then we can write

$$p(s_n, s_{n-1}, \ldots, s_1|H) = p(s_n|H)p(s_{n-1}|H) \cdots p(s_1|H)$$

And using Bayes rule again, we obtain a recursive update formula:

$$p(H|s_n) = \frac{p(s_n|H)p(H|s_{n-1})}{p(s_n|H)p(H|s_{n-1}) + p(s_n|\neg H)p(\neg H|s_{n-1})} \qquad (6.2)$$

---

[b]Not really valid, since readings may relate to the same object, but a commonly made assumption.

In this case $p(s_n|H)$ is given by the sonar model for this cell from the previous subsection, and $p(H|s_{n-1})$ is the prior value in the occupancy grid for this cell.

6.1.3. *Partitioning by Map*

In previous sections we have developed partitioning approaches based on distributing rows, or columns, of values to each processor in the cluster. For this example of building an occupancy grid map from the sonar measurements, we will assume that we are receiving sensor measurements from a team of robots distributed throughout the map. If the robots are not evenly distributed, then some portions of the map, and hence some processors, may be idle. This would be a serious blow to efficiency. We will look at a way to address this assumption presently.

If we have an occupancy grid of size $n \times n$, then we could of course distribute blocks of $(n \times n)/p$ cells (rows and columns are interchangeable in this case) to each processor. Each sensor reading can be written as follows: $sr_s = [(x_s, y_s), (r, \phi)]$, where (x_s, y_s) is the sensor location and (r, ϕ) the range and bearing. Together with the sonar beam angle β and the region sizes from Figure 6-1(a), this can be used to generate a sonar model as in Figure 6-2(b). To determine to which processor(s) to allocate

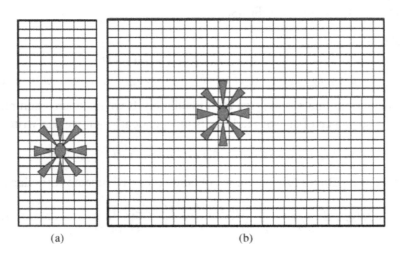

(a)   (b)

Figure 6-3:   (a) Partition of occupancy grid by strips (b) versus checker-board.

the reading, we only need to check whether the map region allocated to the processor intersects the bounding arc of the sonar model $BT(sr_s)$ shown in Figure 6-2(a).

Partitioning the map into long, narrow strips of size $(n \times \frac{n}{p})$ means that the only time a sensor reading will be completely within a single processor's map segment will be when the robot is moving down along the strip (see Figure 6-3(a)). A better choice would be to distribute the map in checker-board segments. These would have to be of an area $(n/\sqrt{p})^2$. In the case where motion is equally likely in all directions, the checker-board would be more likely to have adjacent robot positions on the same block of the map (see (Figure 6-3(b)).

We will assume that initially we know nothing about anything on the occupancy grid; so all values are initialized to 0.5. No data transmission needs to occur to initialize the grid. There are two occasions where data transmission does need to occur:

(1) When sonar readings sr_s are received, it is necessary to determine to which processor(s) to send them.
(2) When the mapping has reached a point where it is necessary to gather the information into a single map, perhaps for transmission to individual robots.

We assume that all m sonar measurements are sent to the root processor from the robot team. Note that measurements might assume that the robot position (as given by (x_s, y_s)) is certain. As we'll see later, because we can capture uncertainty in the robot position by a collection of sample positions, the list of m sonar measurements can include the same range and bearing readings for each possible position of the robot.

We need to associate a 2D coordinate with each processor p which indicates the portion of the occupancy grid BG_p that resides on this processor. For each measurement sr_s the root checks which processor to send the measurement to by intersecting $BT(sr_s)$ with BG_p. In the best case, this is a single processor. In the worst case it will be four processors (if the robot is right at the corner cell of a processor occupancy grid).

We cannot directly use a collective communication operation to gather together the checker-board segments into a single map. The checker-board partition means that, unlike a row or column strip partition, non-sequential chunks of the map are stored on adjacent processors. We will follow a procedure proposed by Quinn [10] for scattering large matrices: Each processor in each row of our 2D processor grid will transmit their portion of the map to the first processor in that row, which then has an n/p strip of the map. Finally, each processor in the first row transmits its strip to the root, which then assembles the entire map.

6.1.4. *Program Design*

The program will have the following overall design:

(1) Phase 0: Initialize the grid topology.
(2) Phase 1: Root traverses the m range readings and allocates each to a processor that has the relevant map section.
(3) Phase 2: Processors read their list of allocated sensor readings and update their copy of the map accordingly.
(4) Phase 3: The two stage map gathering onto the root node is carried out.

Of course steps 2 through 4 could be completed in a loop, continually servicing batches of sonar readings and writing the updated map.

6.1.4.1. *Phase 1*

It is sometimes convenient to view the collection of processors in the cluster as if they were situated on an n-dimensional grid. The MPI_Dims_create function creates a Cartesian grid of processors:

```
int MPI_Dims_create(    int     nnodes,
                        int     ndims,
                        int     *dims    ).
```

The nnodes parameter is the total number of processors that will be in the grid. The ndims parameter is the number of dimensions in the grid — in our case, this will be 2. The result of calling this API function is to generate an array dims which contains the number of processors in each dimension of the grid. In fact, this argument can be an input, to set desired amounts of processors on each dimension. If however it is initialized to zero, then the number of processors in each dimension will be all automatically determined to as close to a square grid as possible.

The following code establishes a 2D grid of processors:

```
int gSize[ 2 ];
gSize[ 0 ] = gSize[ 1 ] = 0;
MPI_Dims_create( p, 2, gSize );
```

where p is the number of processors returned from the MPI_ Comm_size function. It's necessary to have a new communicator to make use of this newly created Cartesian grid processor topology:

```
int MPI_Cart_create (     MPI_Comm     comm_old,
                          int          ndims,
                          int          *dims,
                          int          *periods,
                          int          reorder,
                          MPI_Comm     *comm_cart     ).
```

The MPI_Cart_create function creates a new communicator, comm_cart, to send messages within the Cartesian grid processor topology we have created. The ndims parameter is the number of dimensions and the dims array is the array of processor sizes for each dimension. The array period (set to 1 or 0 for each dimension) controls whether that dimension wraps around. The reorder flag controls whether the processor ranks in the new communicator change from the ranks they had in the old communicator comm_old.

```
Int gPeriods[ 2 ];
MPI_Comm commOccupancyMap;
gPeriods[ 0 ] = gPeriods[ 1 ] = 0;
MPI_Cart_create ( MPI_COMM_WORLD, 2, gSize, gPeriods,
                  0, &commOccupancyMap );
```

The code segment above creates the new communicator commOccupancyMap that we need to implement our partitioning. To determine to which processor to send a reading, we need to know what the map bounds for that processor are. We can use the block macros to implement this in a very straightforward manner:

```
for ( i = 0; i < p; i++ ) /* set up the map boundaries for proc i */ {
        int gPCoords[ 2 ]; /* Processor coordinates in grid topology*/
        MPI_Cart_coords(commOccupancyMap, id, 2, gPCoords);
        mapStart[ i ].x = BLOCK_LOW( gPCoords[ 0 ], gSize[ 0 ], n );
        mapStart[ i ].y = BLOCK_LOW( gPCoords[ 1 ], gSize[ 1 ], n );
        mapEnd[ i ].x  = BLOCK_HIGH( gPCoords[ 0 ], gSize[ 0 ], n );
        mapEnd[ i ].y  = BLOCK_HIGH( gPCoords[ 1 ], gSize[ 1 ], n );
}
```

The function **MPI_Cart_coords** identifies the coordinates of the processor **id** on the grid. The map is partitioned out evenly among all the processors, in a checker-board fashion, based on these coordinates.

6.1.4.2. *Phase 2*

In this phase, the root processor traverses a list of sensor readings, and for each reading **sr[i]**, determines the bounding arc $(BT(s)$ in Figure 6-2(a)) for the sensor reading; this is simplified here to a triangle:

$$\left\{ \begin{bmatrix} x_s \\ y_s \end{bmatrix}, \begin{bmatrix} x_s \\ y_s \end{bmatrix} + R \begin{bmatrix} \cos(\phi + \beta) \\ \sin(\phi + \beta) \end{bmatrix}, \begin{bmatrix} x_s \\ y_s \end{bmatrix} + R \begin{bmatrix} \cos(\phi - \beta) \\ \sin(\phi - \beta) \end{bmatrix} \right\}$$

For each processor j we test if this triangle intersects the rectangle of the **mapStart** and **mapEnd** variables. If it does, we add this reading to a list of items **sr[j]** to send to the processor. When the list of sensor readings has been traversed, and the list for every processor hence completed, the list is sent to the processor with a *point to point send* operation **MPI_Send**:

```
int MPI_Send(        void          *buf,
                     int           count,
                     MPI_Datatype  datatype,
                     int           dest,
                     int           tag,
                     MPI_Comm      comm   ).
```

The address of the data to be sent is given in **buf**, its size in **count** and its type in **datatype**. The rank of the destination processor is in **dest**, and comm is the original, non-grid communicator. The **tag** parameter allows for some selectivity in receiving messages.

```
if ( j != root ) {
        int amount = 4*lsrCount[ j ];
        /* each reading includes [(xs,ys)(range, bearing)] */
        MPI_Send(&amount,1, MPI_INT, j, 1,
                        MPI_COMM_WORLD); /* size of list */
        MPI_Send(&(lsr[ j ]), amount, MPI_FLOAT, j, 1,
                        MPI_COMM_WORLD); /*list */
}
```

The send operating may or may not return immediately (depending on the MPI implementation). However, it relies on the receiving process (see Phase 3) to read the message before it can continue and before it can send another message. For this reason, the root should not try to send a message to itself or it will deadlock!

6.1.4.3. *Phase 3*

Each non-root processor proceeds directly to a *point to point receive* operation. MPI_Recv has similar parameters to the send operation except for one additional argument:

```
int MPI_Recv(   void            *buf,
                int             count,
                MPI_Datatype    datatype,
                int             source,
                int             tag,
                MPI_Comm        comm,
                MPI_Status      *status   ).
```

The additional parameter, status, gives information about the **source** and **tag** arguments of the received message in the case that the receive does not specify specific sender or tag parameters.

The code shown in Figure 6-4 receives the list of lsrCount/4 items from the root and processes them one at a time. The routine **getNext-CellunderSonar** produces each grid cell in turn that falls under the sonar measurement lsr[i]. The probabilistic sonar model (eq. (6.1)) and

```
int gPCoords[ 2 ]; /* Processor coordinates in the grid topology*/
MPI_Cart_coords(commOccupancyMap, id, 2, gPCoords);
int xms = BLOCK_LOW(gPCoords[ 0 ], gSize[ 0 ], n ), /* start of map */
    yms = BLOCK_LOW(gPCoords[ 1 ], gSize[ 1 ], n ),
    xme = BLOCK_HIGH(gPCoords[ 0 ], gSize[ 0 ], n ), /* end of map */
    yme = BLOCK_HIGH(gPCoords[ 1 ], gSize[ 1 ], n );

MPI_Recv( &lsrCount, 1, MPI_INT, root, 1, MPI_COMM_WORLD,
                &status);   /* get amount */
MPI_Recv( &lsr, lsrCount, MPI_FLOAT, root, 1, MPI_COMM_WORLD,
                &status);   /* get readings */

for ( i = 0; i < lsrCount/4; i++ ) { /* process each sonar reading*/
    int x = y = -1; /* coordinates of cell under sonar */
    while ( getNextCellunderSonar( lsr[ i ], &x, &y) != 0 )
        if ( onProcessorMap(x, y, xms, yms, xme, yme) ) {
            /* do Bayesian update */
            sLikli = sonarModel( lsr[ i ], x, y );
            sProb = map[ x-xms][ y-yms ] * sLikli ;
            map[ x-xms ][ y-yms ] = sProb / ( sProb +
                        (1 – map[ x-xms ][ y-yms ])*(1 - sLikli ) );
        }
}
```

Figure 6-4:   Occupancy grid update.

the Bayesian update (eq. (6.2)) formulas are then calculated and used to update the map. It is necessary to offset the coordinates of the map by (xms, yms) — the start coordinates of the portion of the map on each processor — transforming the indices so they can be used on the processor's section of the map which starts at (xms, yms).

6.1.4.4.   *Phase 4*

The final phase involves transmitting the map back to the root for redistribution to the team of robots or to a command center for viewing. We use a gather command to pull the map together and this command also serves to synchronize all processors, since different processors may finish processing their readings at different times. We first need to allow collective communication operations on just the subgroup of processes that are each in a row of the map. The MPI_Comm_split function allows us to subdivide the processors in a communicator:

```
int MPI_Comm_split (    MPI_Comm    comm,
                        int         partition,
                        int         rank,
                        MPI_Comm    *comm_out    ).
```

We want to have each row of the map grid be a communicator with the zero element of the row as the root processor. Each processor executes this command, providing its coordinates on the processor grid gPCoords. We use the first dimension (the row number) as the partition, and we will use the second dimension as the rank within the partition.

The code in Figure 6-5 shows the two stage gathering process. The first stage accumulates strips of the map of size $n \times \sqrt{p}$ in each of the processors of the first row, that is, the root processors of the row communicator. These are sent by scanning the map row by row. Once the strip has been built up on each processor we use the technique shown in Chapter 4 to convert between a 1D and a 2D array. We keep the same notation here: mapRGdata (the strip map collected on the row roots) is the 1D array for the 2D mapRG array, and mapGdata (the full map) is the 1D array for the 2D mapG. We establish communicators for the columns in the same way we did for the rows. However, the final gather is executed *only* by the processors in the zero column (the row roots). This is a single gather using the 1D array addresses and it builds up the final map on the root process.

6.1.5. *Analysis*

The computation time for the map update is a function of the number of sensor measurements. Our main assumption is that the robot team members remain distributed so that the readings are distributed roughly evenly among processors. If the time to execute the 'process sonar' loop in Figure 6-4 is χ then the serial time to process all m sensor readings is $T_s = m\chi$. If the readings are distributed equally, then the parallel calculation time is $T_{cal} = m\chi/p$. However, we incurred parallel overhead as follows:

(1) *Partitioning readings*: Each reading had to be checked for intersection with each processors grid. If the intersection cost was ζ then in the

```
/* stage one, send to row root */
MPI_Comm commOccupancyRow;
/* row communicator for each processor*/
MPI_Comm_split (commOccupancyMap, gPCoords[ 0 ],
                gPCoords[ 1 ], &commOccupancyRow);

int stripWidth = BLOCK_SIZE(gPCoords[ 1 ], gSize[ 1 ], n );

for ( i = 0; i< BLOCK_SIZE(gPCoords[ 0 ], gSize[ 0 ], n ); i++)
    MPI_Gather(&map[ i ][ 0 ], stripWidth, MPI_FLOAT,
            /* local n/√p times n/√p map */
            &mapRG[ i ][ 0 ], stripWidth, MPI_FLOAT,
            /* map stripe n/√p times n */
            commOccupancyRow );

/* stage two, root column sends to root */
MPI_Comm commOccupancyCol;
/* row communicator for each processor*/
MPI_Comm_split (commOccupancyMap, gPCoords[ 1 ],
                gPCoords[ 0 ], &commOccupancyRow );

if (gPCoords[ 1 ] == 0) /* only row root processors do this*
    MPI_Gather(&mapRGdata, n* stripWidth, MPI_FLOAT,
            /* map stripe n/√p times n */
            &mapGdata, n*stripWidth, MPI_FLOAT,
            /* fullmap n times n */
            commOccupancyCol);
```

Figure 6-5:   Gathering occupancy grid on root processor.

worst case we have a processing cost of $mp\zeta$ and a communication cost of $4m(\lambda + 4/\beta)$.

(2) *Gathering the array, Stage 1*: The first stage requires $n/\sqrt{p}$ gather operations, each on data of size $n/\sqrt{p}$. Let's simplify this to one gather of size n^2/p. Although all p processors are involved, the rows do their operations in parallel with each other, taking a time that is a function of $\sqrt{p}$ each. Using our collective communication model, this is $\lambda \log \sqrt{p} + n^2(\sqrt{p} - 1)/\beta p$.

(3) *Gathering the array, Stage 2*: The second stage requires one gather operation of size $n^2/\sqrt{p}$ by $\sqrt{p}$ processors. Again, using our collective communication model, this is $\lambda \log \sqrt{p} + n^2(\sqrt{p} - 1)/\beta\sqrt{p}$.

Adding and rearranging, the total parallel time is:

$$T = m \left[\frac{\chi}{p} + p\zeta \right] + \left[\lambda \log p + \frac{n^2(p-1)}{p\beta} + 4m \left(\lambda + \frac{4}{\beta} \right) \right]$$

If we could have done a single gather, the time would have been the same. The simplification we made in item 2 (assuming one gather operation rather than $n/\sqrt{p}$ gather operations) means our actual communication time will be worse than this of course.

6.1.6. *Partitioning by Sensor Readings*

Our strongest assumption in the previous approach was that sensor readings were distributed over the map. We have no control over this and, in the worst case, all the sensor readings might be in just one processor's portion of the map, resulting in $p - 1$ idle processors! An approach that addresses this issue is to consider the map to be resident in full on every processor and then partition the sensor readings equally, sending m/p to every processor. But in that case, it's entirely possible that several processors may end up working on the same portion of the map, and we then have the issue of how to combine these into a single map.

Fusion of maps from two robots (e.g., [1]) requires understanding the spatial transformation from one robot's map to the other, that is, the issue of *map alignment*. In our case, we have all the robots represented on a single map. The uncertainty in robot position with respect to this common map is captured by uncertainty in the sensor location when the readings are processed onto the map. Therefore there is no issue of map alignment to be solved.

Each processor is producing a map of evidence values based on the sensor readings that were allocated to it. Map fusion is the process of fusing the evidence values from each processor onto a single map. This is similar to the fusion of occupancy maps produced by different sensors with a known spatial relationship, e.g., as described in [14], where the maps are fused together using a weighted linear summation technique. We adopt this approach, and we use the number of readings that cover each grid cell as a way to determine the weight factor in the summation. When fusing a map cell from cell values on two or more processors, the values that were the result of more readings will have a higher weight than those from fewer (or no) readings.

6.1.7. *Program Design*

This partitioned sensor readings program will differ from the previous partitioned map program in distributing sensor readings and in fusing the resultant maps. The map update procedure will, however, be the same as the 'for loop' in Figure 6-4.

The distribution of sensor readings can be carried out with a single scatter function, distributing m/p readings to each processor's local sensor reading list *lsr*. Each processor has a full copy of the map, *mapG*, with $n \times n$ values.

Once the sensor readings have been used to update each processor's copy of the map, it is necessary to fuse the local maps back onto the root processor for distribution to the robot team or for viewing in a command center. To implement the proposed map fusion approach, we add a second 2D array to keep track of the number of readings that covered each cell, *countG*. Whenever we calculate a Bayesian update to a cell, we will increment the *countG* value for that cell. When all updates are done, we multiply the two arrays on an element-wise basis, and carry out a reduction operation for each element:

```
for ( i = 0; i < n; i++ )
        for ( j = 0; j <n; j++ ) /* mul. map by local weight array */
            mapG[ i ][ j ] *= countG[ i ][ j ];
MPI_Reduce(mapG, mapG_r, n*n, MPI_FLOAT, MPI_SUM,
            root, MPI_COMM_WORLD);
MPI_Reduce(countG,countG_r, n*n, MPI_FLOAT, MPI_SUM,
            root, MPI_COMM_WORLD);
if (id == root) {
        for ( i = 0; i < n; i++ )
            for ( j = 0; j < n; j++ ) /* normalize map*/
                mapG_r[ i ][ j ] /= countG_r[ i ][ j ];
}
```

The weighted, reduced values need to be normalized: each cell is divided by the sum of all updates from every processor on that cell. For that reason, we need to also reduce the countG array, to get the sum of all updates for each cell. The root then divides the reduced mapG_r array by the reduced countG_r weight values on an element by element basis.

6.1.8. *Analysis*

The serial time for this algorithm is the same as for the map partition algorithm, $T_s = m\chi$. The parallel calculation time for the algorithm is basically the same also, but with a fixed extra cost for the map fusion. This cost includes updating each of the n^2 countG cells and weighting each of the n^2 map values by the corresponding count: $n^2\zeta$.

The parallel overhead comes from three collective communication operations:

(1) Scattering the readings: Scattering $4m/p$ values to each processor requires a time $\lambda \log p + 4m(p-1)/p\beta$.
(2) Reducing the local maps: This is two reductions of n^2 values, requiring a time $2(\lambda + n^2/\beta) \log p$.

The total time is therefore:

$$T = \left[\frac{m\chi}{p} + n^2\zeta\right] + \left[3\lambda \log p + \frac{n^2}{\beta}\log p + \frac{4m(p-1)}{p\beta}\right]$$

Comparing this with the map partitioning approach:

(1) While the parallel calculation time is basically the same for both, it is only valid for the map partition algorithm if robots are distributed evenly over the map, where the sensor reading partition algorithm enforces this division of labor.
(2) The map fusion incurs a much higher, fixed (by the map size) communication overhead than the map gathering operation,

6.2. Monte-Carlo Localization

In Chapter 4 we introduced the concept of dead-reckoning: using the information about what motion commands a robot has carried out to determine where the robot is. When a robot attempts to carry out a motion command, the actual motion produced by the platform may differ from the commanded motion for a number of reasons, including backlash in the gearing and loss of traction between the wheels and ground. In a robot such as the Pioneer 3-AT, changes in orientation in particular are difficult to model with dead-reckoning. The 3-AT rotates by 'skid steering' — that is, when the left wheels and the right wheels have different rotational velocity, the platform rotates roughly around its center as a result of wheel skidding. On uneven terrain, skid steering produces extremely variable results.

Localization is the process of determining where a robot is with respect to a map. There are several approaches to localization, including Kalman Filtering [3] and Markov grids [11]. Both of these approaches represent the uncertainty about the pose of the robot by means of a parametric distribution model. The technique we utilize here is based on so-called *Monte Carlo*[c] methods. Monte Carlo methods use repeated statistical sampling as the basis for calculating some quantity. Monte-Carlo Localization (MCL), introduced by Thrun *et al.* [13], represents the uncertainty about the pose of the robot by a set of samples (sometimes called particles) of the posterior distribution of robot poses. Because it represents the uncertainty by a collection of weighted samples, MCL does not suffer from the same constraints on the parametric form of the distribution that other approaches do. But furthermore, MCL turns out to be straightforward to implement and it has become quite widely used.

The MCL algorithm starts at time step $t - 1$ with a collection of n particles S_{t-1}. These represent the possible poses of the robot in this time step. Each particle is propagated forward in time based on the commanded motion of the robot at step t by using the *motion model* for the robot. The new collection S'_t represents all the locations that the robot could have ended up in starting from S_{t-1}. The sensor readings from the robot are now taken into account, and the samples weighted by how well they agree with the observed sensor readings. This is accomplished using the *perceptual model* for the senor. Finally, S'_t is *resampled* to produce a new collection of n samples, S_t in such a manner that the higher weighted samples are more

---

[c]Named after the (in)famous gambling city in the principality of Monaco in the south of Europe.

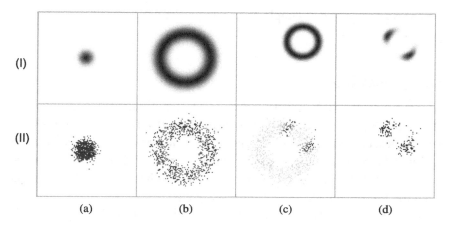

Figure 6-6:   Stages in Monte Carlo Localization (from [4]).

likely to be selected for the new collection. Figure 6-6 shows a typical example of the evolution of the set of particles from S_{t-1} through S_t.

In Figure 6-6, the top row, labeled (I), illustrates the exact probability distribution, whereas the bottom row, labeled (II), illustrates the collection of samples. Figure 6-6(a) shows S_{t-1}. The position of the robot is somewhat localized but the orientation is completely unknown. A move of $1\,m$ is commanded, and the resulting collection S'_t is shown in Figure 6-6(b). Sensor readings indicate an object in the top right corner (Figure 6-6(I)(c)) and the particles are weighted according to this (Figure 6-6(II)(c); higher scoring particles show darker). Finally, the weighted collection is resampled such that higher scoring particles are more likely to be chosen, S_t shown in Figure 6-6(d). This final step selects mostly high scoring particles, but some lower scoring particles can be selected as well. This selection procedure helps avoid local maxima: some of the lower-scoring particles may end with much higher scores in later iterations.

The motion model for the robot is captured as a conditional probability density $p(x'|x, u)$ that the robot is at x' given that it was at x and executed a motion command u. A sampling motion model takes inputs x and u, and generates an x' according to the distribution $p(x'|x, u)$.

The perceptual model is the conditional probability density $p(o|x)$, the probability of the reading o given the robot is at x. Given an observed sensor reading o and a sample at x, this probability is used to weight the sample.

6.2.1. *Partitioning*

Monte-Carlo methods translate well to cluster implementations because they require little inter-processor coordination [10]. In this case, if we have m samples, each sample can pass through the phases of motion propagation and weighting independent of every other sample. The partitioning in this case is quite straightforward: each processor will handle m/p samples and send the result back to the root for resampling.

This extends easily to the case where there are r robots, as long as it is acceptable to process all the robot positions and sensory measurements in synchronized steps. Each step includes a motion propagation, a weighting and a resampling, as discussed before. For example, if the motions and measurements of the robot team are time-stamped when sent to the cluster, they can be grouped into intervals. All the motions and measurements that fall in a single interval can be considered concurrent and can be processed in the same step.

6.2.2. *Program Design*

The MPI program below is developed for the single robot case. We will assume that a map mapG has already been broadcast to all processors.

- The set of m particles will be held in a $4m$ array samples, representing the position and orientation of each particle and also a weight value. The ith particle starts at the $4i^{th}$ array location. The array is initialized to random valid locations and orientations for the robot on the map (i.e., not inside a piece of furniture or a wall).
- The motions of the robot will be represented by an $n \times 2$ array motion of rotation and forward distance values, the sequence of motion steps that the robot is commanded to carry out.
- The sensory information from the robot will be represented by an $n \times n_s$ array range where n_s is the number of range readings for each motion step (for example, 8 sonar readings or 180 laser readings).

The program will loop through the n motion commands and sensor measurements and carry out the motion propagation, weighting and resampling for each motion command. This step provides the opportunity for parallelism and we will look at it in more detail.

The coordination of the parallelism will be carried out by the root processor. The root begins by scattering the m particles to p processors.

```
MPI_Scatterv( &samples[ 0 ],   displ, dendcnt, MPI_FLOAT,
              &samples_l[ 0 ], dendcnt[ rank ], MPI_FLOAT,
              0, MPI_COMM_WORLD );
```

Here samples_l is the set of particles allocated to the processor. To scatter the samples array as discussed in Chapter 5, the arrays of displacements (displ) and sizes (dendcnt) are calculated. These are based on the number of particles (not the size of the array — that might lead to a particle being split over two processors!).

The root then broadcasts the motion command and sensory measurements for this step (where i is the step index).

```
MPI_Bcast ( motion[ i ], 3, MPI_FLOAT, 0, MPI_COMM_WORLD );
MPI_Bcast ( range[ i ], n_s, MPI_FLOAT, 0, MPI_COMM_WORLD );
```

Each processor now has sufficient information to carry out the motion propagation and weighting. The motion propagation produces a new position for each particle, where the position is a single sample from the motion model given the commanded motion,

```
motionModel( samples_l[ s ], motion[ i ] ).
```

To weight this particle based on the sensory measurements, it's necessary to compare each range reading with the predicted range reading from the particle position. By using ray tracing [15] from the particle location on the map, mapG, a predicted range reading for the particle can be calculated. The probability model from Figure 6-2(b) can then be used to determine the probability of getting the measurement in range[i] for this sensor given the predicted range. This probability is used as the weight of the particle. The weighted particles are then returned to the root.

```
for (s=0; s< dendcnt[rank]; s++) { /* traverse samples */
    motionModel( samples_l[ s ],motion[ i ] );
    weightSample( samples_l[ s ],range[ i ], mapG );
}
MPI_Gatherv(&samples_l[ 0 ], dendcnt[ rank ], MPI_FLOAT,
            &samples[ 0 ], displ, dendcnt, MPI_FLOAT,
            0, MPI_COMM_WORLD );
```

The resampling step can be implemented on the root processor by first sorting the particles according to their weight. The cumulative weight for each particle is calculated (the sum of the particle weight plus all lower weighted sample weights). A particle is selected by generating a random number between 0 and the sum of all weights and by finding the particle with the closest but higher cumulative weight. The cumulative reweighting ensures that samples with higher weight will be chosen with higher probability.

It may be more convenient to represent a particle as a C *struct* rather than a 1-D array. This adds a little more complication to the message-passing. MPI conducts type checking during the message-passing process and hence needs to be explicitly told the layout and types of the *struct* members. It is possible on a homogeneous cluster to transmit and receive a *struct* s as a block of data of size *sizeof(s)* and of type MPI_BYTE. However, this completely bypasses the type checking in MPI and introduces non-portability issues in the program.

6.2.3. *Analysis*

Let us consider the time for one processor to carry out the motion propagation and weighting on one sample to be χ. The sorting and resampling, if done efficiently, will take an additional $m \log m$ and the steps need to be repeated for all motions:

$$T_s = nm(\chi + \log m)$$

The parallel calculation time per step (assuming p divides m for simplicity) is $m\chi/p$ for all processors, but then the root additionally needs to do the sorting and resampling, $T_{cal} = nm(\chi/p + \log m)$. The communication time per step includes the broadcast of $n_s + 3$ values and

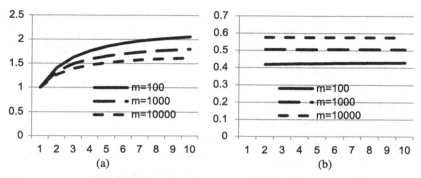

Figure 6-7: Parallel Monte-Carlo Localization ($\lambda = 3\,\mu s$, $\beta = 100\,Mbps$, $n_s = 180$, $\chi = 7.3\,\mu s$) (a) Graph of Speedup and (b) Graph of Karp-Flatt metric versus number of processors.

the scatter/gather of $4m/p$ values. The total parallel time is then:

$$T_p = nm \left(\frac{\chi}{p} + \log m \right) + n \left[k \log p + \frac{8m(p-1)}{p\beta} \right]$$

where $k = 3\lambda + (n_s + 3)/\beta$ a term that depends on the number of range sensors used in localization. Figure 6-7(a) shows a graph of the speedup $\psi = T_s/T_p$ for this algorithm. While the parallel implementation clearly has timing advantages over the serial implementation, it is interesting to note that the speedup is quite poor and decreases for larger problem sizes (larger m) in this implementation.

The experimentally determined serial fraction of the code, the Karp-Flatt metric defined in Chapter 3 can yield some insight here. Figure 6-7(b) shows a graph of the Karp-Flatt metric — which is increasing slowly with problem size. To get better performance we need to address the serial fraction of the code.

6.2.4. *Improving the Serial Fraction*

There are two operations in the serial fraction: the sorting of the particle list and the resampling of the particle list.

The most efficient sequential sorting of a list of m particles requires a time $m \log m$. Sorting can be done in parallel with a time of $\frac{m}{p}(\log m + \log p)$ [10] and one option is to replace the section of the program where the root doing all the sorting with a parallel sorting phase. This option especially makes sense if there is a parallel sort library available.

We can also make a small modification to the code to carry out the sorting in two steps: a local sort will be carried out on each processor, and the local lists will just be merged on the root processor. Since the local sorts can be done in parallel with each other, their time is $\frac{m}{p} \log \frac{m}{p}$. The root then has to merge the p sorted lists, which is a linear operation in the size of the lists, $p \times \frac{m}{p} + 1$.

Resampling of the m particles is carried out as follows: A particle is selected by generating a random number between 0 and the sum of all weights and by finding the particle with the closest but higher cumulative weight. Since the list of particles is ordered, a binary search will take $\log m$. However, the search must be repeated for each particle, for a total time of $m \log m$. This is the same time as the sequential sort. Hence, if we don't address this as well, then our improvement to the sorting time will be of no benefit.

The resampling operation for each particle is independent and could be carried out in parallel in blocks of m/p on p processors. This approach would require the following steps:

(1) Broadcast the list of particles.
(2) Each processor resamples its block of m/p particles.
(3) Gather the resampled list on the root.

However, note that the first step in the MCL loop after the resample is to scatter the particles again! The gather in step (3) can be eliminated as can the scatter. The only sequential section of the code is now the (linear time) merging of the sorted sublists and the calculation of the cumulative scores. The revised time, omitting the scatter but including the resample broadcast as well as the sublist sorting, merging and sublist resampling, is as follows:

$$T_p = \frac{nm}{p} \left(\chi + 2 \log \frac{m}{p} \right) + n(m + 1) + n \left[k' \log p + \frac{4m(p - 1)}{p\beta} \right]$$

where k' is now includes the resampling broadcast time:

$$k' = 3\lambda + \frac{n_s + 3 + m}{\beta}$$

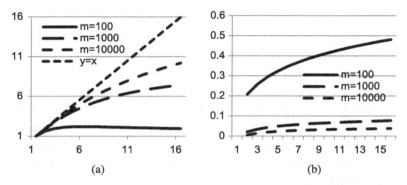

Figure 6-8: Parallel Monte-Carlo Localization with improved serial fraction ($\lambda = 3\,\mu s$, $\beta = 100\,Mbps$, $n_s = 180$, $\chi = 7.3\,\mu s$) (a) Graph of theoretical Speedup and (b) Graph of theoretical Karp-Flatt metric versus number of processors.

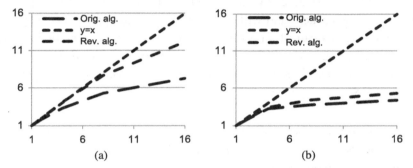

Figure 6-9: Comparison of original parallel Monte-Carlo Localization algorithm with improved serial fraction algorithm; (a) Graph of Speedups for $m = 10000$ and (b) for $m = 100$ versus number of processors, measured on a cluster on an HPC cluster of quad-core dual processor Intel Xeon® processors.

Figures 6-8(a) and 6-8(b) show the estimated speedup and serial fraction calculations for the improved serial fraction using the theoretical performance models. The Karp-Flatt metric now shows that the experimentally determined serial fraction falls for larger problem sizes. This is reflected in a dramatically improved (predicted) speedup.

Figures 6-9(a) and 6-9(b) show this speedup as measured on an HPC cluster of quad-core dual processor Intel Xeon® processors. The improvement in speedup is clearly visible. This shows the dramatic increase for larger problem sizes. The poor speedup predicted for smaller problem sizes is also clearly visible.

6.3. Summary

In this section we have looked at two important robotics problems: building a map, and given a map, determining where the robot is on the map. We looked at mapping in the context of robot teams returning sensory information to a common source where that information is fused into an occupancy map. The initial implementation assumed a team of robots distributed over the map, and partitioned the map in checkerboard segments among the team. Our second approach placed a full copy of the map at each processor and distributed the sensor readings. This is by far the more robust approach in general.

Occupancy maps do not scale that well. One option is to use dynamically expanding occupancy grids [7] which add additional cells as necessary. Another alternative is to combine the representation with a topological mapping approach (and will discuss this approach presently) [11].

Monte-Carlo localization is a nice fit for parallel implementation as long as the resampling does not force a sequential component (as we saw with sorting). A team of robots can be accommodated easily by this algorithm in two ways:

(1) Processors can be partitioned among the robot team. If there are r members of the team, then the parallel MCL is carried out for each robot on r/p processors.
(2) Alternatively, sample, motion and sensory measurement data from multiple robots can be sent to each processor.

For contrast, see the parallel implementation of mapping and localization described in [9] which is based on Thrun's [12] probabilistic concurrent mapping and localization rather than MCL.

References

1. Andersson, L.A.A. and Nygards, J., On multi-robot map fusion by inter-robot observations, *12th International Conference on Information Fusion (FUSION '09)*, Seattle WA, 2009.
2. Arbib, M.A., *The Metaphorical Brain 2*, Wiley & Sons 1989.
3. Choset, H., Lynch, K., Hutchinson, S., Kantor, G., Burgard, W., Kavraki, L., and Thrun, S., *Principles of Robot Motion*, MIT Press 2005.

4. Dellaert F., Fox, D., Burgard W., and Thrun, S., Monte Carlo Localization for Mobile Robots, *IEEE International Conference on Robotics and Automation (ICRA99)*, Detroit MI, May 1999.

5. Dudek, G. and Jenkin, M., *Computational Principles of Mobile Robotics*, Cambridge Press, 2000.

6. Elfes, A., Using Occupancy Grids for Mobile Robot Perception and Navigation, *IEEE Computer*, June 1989.

7. Ellore, B.K., Dynamically expanding occupancy grids, *Master's thesis, Texas Tech University*, 2002.

8. Moravec, H. and Elfes, A., High Resolution Maps from Wide Angle Sonar, *IEEE International Conference on Robotics and Automation*, Washington DC, 1985.

9. Lyons, D.M. and Isner, G.R., Evaluation of a Parallel Algorithm and Architecture for Mapping and Localization, *7th International Symposium on Computational Intelligence In Robotics and Automation, CIRA 2007*, Jacksonville FL June 20–23, 2007.

10. Quinn, M.J., *Parallel Programming in C with MPI and OpenMP*, McGraw-Hill 2004.

11. Thrun. S., *Robotic mapping: A survey*. In G. Lakemeyer and B. Nebel, editors, *Exploring Artificial Intelligence in the New Millenium*, Morgan Kaufmann, 2002.

12. Thrun, S., Burgard, W., and Fox, D., A Probabilistic Approach to Concurrent Mapping and Localization for Mobile Robots, *Machine Learning and Aut. Robots* 31/5 1998, pp. 1–25.

13. Thrun, S., Fox, D., Burgard, W., and Dellaert, F., Robust monte carlo localization for mobile robots, *Artificial Intelligence* 128(1–2): 99–141, 2001.

14. Usher, K., Obstacle avoidance for a non-holonomic vehicle using occupancy grids, *Proceedings of the 2006 Australasian Conference on Robotics & Automation*, Auckland, New Zealand, December 2006.

15. Wilkes, D., Dudek, G., Jenkin, M., and Milios, E., Modeling Sonar Range Sensors, in: (Archibald and Petriu, Eds) *Advances in Machine Vision: Strategies and Applications*, World Scientific Press, Singapore 1992.

Chapter 7

Vision and Tracking

The sense of vision is a powerful and dominating sense in humans. As Dudek *et al.* [7] note: Anyone who doubts the importance of vision to human navigation should consider the problems faced when walking around in the dark! Since people program robots and, many times, program them to operate in the same environment, with the same tools that people themselves use, it is natural to consider equipping a robot with a visual sensor. A typical approach is to use a video camera as a visual sensor. The subjective experience of understanding the content of a visual scene is often as simple as opening one's eyes — there seems to be little or no conscious effort involved in perceiving a three-dimensional world filled with objects and surfaces both near and far.

The perception is so effortless that it is possible, even typical, to underestimate the difficulty of duplicating the feat with a video camera. The lack of effort also belies the amount of neural hardware that goes into this visual illusion [1]. Extracting information from a video sequence is a computationally intensive and difficult problem. While the field of computer vision continues to make important advances, it would be premature to consider it a problem solved.

We first briefly look at the issues that make computer vision difficult. Figure 7-1(a) shows *the pin-hole camera model*. A pin-hole camera is a closed container with a small hole on one side. Light rays enter through the hole, illuminating the back wall with the color of the object from which they were reflected and thereby painting an image on the wall.

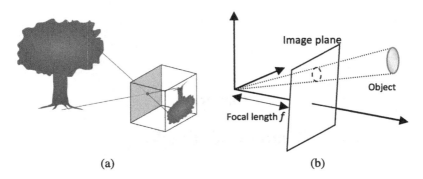

(a)   (b)

Figure 7-1:   Pin-hole camera (a) and its mathematical model (b).

The image is upside-side down and reversed because of the path the light rays travel. This relationship between the object and its image is often modeled mathematically as a *perspective projection* as shown in Figure 7-1(b). The image plane in Figure 7-1(b) is shown with a coordinate system aligned and centered on the camera axis. The image plane is located a distance f from the origin, which is the optical center of the camera. Let the coordinates of a point on the image plane be $[u, v]^T$. Since the image plane is positioned at distance f from the center of projection, the three dimensional coordinates of that point are $[u, v, f]^T$. Following a line from the origin through $[u, v, f]^T$ and out to the object, call the coordinates of the point that the ray intersects on the object $[X, Y, Z]^T$. The relationship of the image coordinates and object coordinates can be written:

$$u = f\left(\frac{X}{Z}\right), \quad v = f\left(\frac{Y}{Z}\right)$$

Information is lost in the 3D to 2D transformation. This is the first problem that makes computer vision difficult: the 2D structure seen on the image is related in a complex way to the 3D structure of the original scene.

An image is represented on a computer as an array of pixels. Each pixel corresponds to the intensity and the color of light integrated over a small area of the image. A pixel can be represented as a single value, corresponding to the total intensity of light in that area. It can also be represented as a n-tuple of values in some *color space*. A typical color space is RGB (Red-Green-Blue). A large variety of colors can be

Figure 7-2: (a) Close up (b) of a section of a digital image.

represented by a linear combination of the 'standard' red, green and blue primary colors[a] — the coefficients in this linear combination are the pixel's RGB values. These will often be numbers between 0 and 1. For example: the RGB value $(1, 0, 0)$ would denote a pixel that is 100% red; $(0.25, 0.25, 0)$ would be a pixel with 25% red and 25% green, a dark yellow; $(0, 0, 0)$ would be black; and $(1, 1, 1)$ would be white.

Figure 7-2 illustrates a second difficulty in computer vision. To a human, a scene such as Figure 7-2(b) consists of some objects — mountains, trees, river — positioned in the scene relative to each other. To a computer, the image has to be understood pixel by pixel, as in Figure 7-2(a) and the scene viewed at this scale is far less clearly separable into regions corresponding to objects.

7.1. Following the Road

Consider the problem illustrated in Figure 7-3. This figure shows a robot using its visual sensor to navigate along a path delineated by white lines on the ground. The problems involved are to first identify the lanes (the white lines in the image) and then determine where these lines are in space with respect to the robot. We will look at some of the computational

---

[a] SMPTE, the Society of Motion Picture and Television Engineers, among others, has established standard red, green and blue primaries.

| (a) | (b) | (c) |

Figure 7-3:   (a) Robot lane following (b) lane closeup and (c) thresholded Cb component of closeup.

steps in road following and how this might be done on a cluster. A robot such as the Pioneer 3-AT shown in Figure 7-3(a) does not typically carry an onboard cluster. However, it is not unreasonable to expect that such machines will carry multiprocessor, multicore computers in the near future. More sophisticated robots, such as Junior, the Stanford University entry in the DARPA Urban Challenge already carry multiprocessor, multicore systems [15] onboard.

7.2.   Iconic Image Processing

Before we can consider identifying image regions corresponding to objects in the scene, we will typically need to preprocess the image at the pixel level. The noise introduced into the image by the camera or random lighting effects may need to be reduced. Aspects of the image such as edges or regions of color may need to be enhanced. These operations consider the image only to be a collection of pixels, and operate pixel by pixel, (or pixel by small regions of pixels). This kind of processing is called *iconic image processing* to distinguish it from operations that attempt to identify 3D structure, objects or shapes in the image.

Iconic image processing is typically more suited to fine grain, massively parallel, single instruction multiple data architectures [4]. However, as long as the image information can be moved efficiently enough, cluster processing of images can also be effective. One approach to avoiding image transfer costs is to equip individual nodes in the cluster with separate cameras [8].

Examples of pixel-based iconic image operation include color space transformations, color filtering, intensity transformations and so forth. For example, the difference between the white lines and the grass background in Figure 7-3(a) becomes much more apparent when the color space is changed from RGB to the luma-chroma colors space *YCrCb*, subsequently restricted to just the Cb chroma component, and finally thresholded to a binary image. *Thresholding* is also a pixel-based iconic image processing operation. Pixels in the thresholded binary image are set to 1 if the pixels in the original image have a value greater than the *threshold* value and are set to 0 otherwise.

Changes in the lighting of a natural scene are typically not very noticeable to a human observer, but can cause a great deal of difficulty in analysis of images. A scene might have overall increases or decreases in light intensity due to the time of day or weather conditions. There may also be shadow effects as lighting sources move due to the time of day, or as objects move in or adjacent to the scene. Another reason to use YCrCb, rather than RGB, for the analysis of natural images (such as Figure 7-3(a)) is that the CrCb chroma (or color) components are less sensitive to changes in light intensity. The scene lighting intensity variations are mostly confined to the Y luma component. In RGB all three components can vary with intensity. Hence, the Cb component is a more reliable 'white line' detector over a wide range of lighting conditions.

The formulae to change RGB to YCrCb (from the CCIR-601 digital video standard) are:

$$Y = 0.299R + 0.587G + 0.114B$$
$$Cr = (R - Y) * 0.713 + 128$$
$$Cb = (B - Y) * 0.564 + 128$$

Notice that there are quite a few isolated white speckles in the image in Figure 7-3(c). These could be reduced by reducing the threshold value. However, if we do that, we may also reduce the amount of the white line that is visible. There is a better solution, a different class of iconic image processing that we will describe presently.

7.2.1. *Partitioning*

In this book so far, we have looked at several ways to partition matrices of data among cluster nodes: we can distribute strips by row and by column as we did in Chapter 5 or we can distribute by checkerboard blocks as we did in Chapter 6. From the perspective of a pixel-based iconic image operation, where all the pixels are operated upon independently of each other, none of these partitioning approaches has an advantage over the other in terms of communication costs. If the number of processors available is greater than the row size or column size of the image, then a checkerboard distribution is the better choice since it allows all the processors to be used. If the image is stored in consecutive rows in memory, then block distribution by row is the best choice since it requires no additional processing to set up a scatter operation.

7.2.2. *Program Design*

The root processor distributes the $n \times m$ color image to all processors by row. The block macros are used to allocate strips of the image for scattering using MPI_Scatterv to each processor. The arrays displ and dendcnt will contain the size and start location of each processor's block of memory in the image array. These are initialized using the block macros so that each processor is allocated a share of full rows. Since there are 3 color pixel values in m columns, there are $3\,m$ values in each row. We will assume each color component is one byte long (this is also referred to as 24-bit color).

```
for ( i = 0; i < p; i++ ) {
        displ[ i ]   = BLOCK_LOW(id, p, n)*3*m;
        dendcnt[ i ] = BLOCK_SIZE(id, p, n)*3*m;
}
```

The image can then be scattered to each processor as follows:

```
MPI_Scatterv( &image, displ, dendcnt, MPI_BYTE,
              &image_l, dendcnt[rank], MPI_BYTE,
              0, MPI_COMM_WORLD );
```

Each processor will see its portion of the image in image_l and will carry out its color conversion and thresholding on that block, producing a thresholded output image, result_l. The output image has just byte per pixel (and in fact could be compressed even more to just 1 bit per pixel).

```
for ( y = 0; y < BLOCK_SIZE(id, p, n); y++) /* row index */
    for ( x = 0; x < 3*m; x++ ) /* column index */ {
        /* color convert the image RGB pixel values*/
        index=y*3*m+x; indexR = y*m+x; /* index into images */
        Y = 0.299*image_l[index ] + 0.587 * image_l[index +1 ]
                                   + 0.114 * image_l[index +2 ];
        Cr = (image_l[index]    − Y)*0.713 + 128;  /* not used */
        Cb = (image_l[index +2] − Y)*0.564 + 128;
        /* Threshold based on the Cb component*/
        if (Cb>Threshold) result_l[indexR]=1;
        else result_l[indexR ]=0;
    }
```

The result image is gathered back onto the root. However the block sizes and displacements need to be modified, since the thresholded result image is *three times smaller* than the original color image.

```
for ( i = 0; i < p; i++ ) {
    displ[ i ] /=3;
    dendcnt[ i ] /= 3;
}

MPI_Gatherv( &result_l, dendcnt[rank], MPI_BYTE, &result,
             dendcnt, displ, MPI_BYTE, 0, MPI_COMM_WORLD );
```

7.2.3. *Analysis*

Let the time taken to color convert and threshold 1 pixel be χ. In that case, the serial time to process an $n \times m$ image is easily calculated as $T_s = nm\chi$.

The parallel communication time is composed of the time to scatter the color image strips and to gather the thresholded image back again. The time to scatter the image strips from the root to each processor is given by $\lambda \log p + 3nm(p - 1)/p\beta$. The time to gather the result image back to the root is $\lambda \log p + nm(p - 1)/p\beta$. The parallel calculation time is $nm\chi/p$.

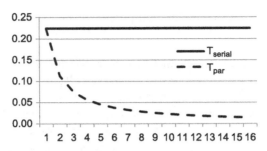

Figure 7-4:   Performance graph for parallel pixel-based operations for 640×480 image ($\lambda = 3\mu s$, $\beta = 1\,\text{Gbps}$, $\chi = 730\,\text{ns}$).

(a)   (b)   (c)

Figure 7-5:   (a) Original image; (b) smoothed; (c) edges.

The total parallel time is therefore

$$T_p = 2\lambda \log p + \frac{4nm(p-1)}{p\beta} + \frac{nm\chi}{p}$$

7.2.4.   *Spatial Pixel Operations*

There is an important class of pixel-based operations that involves looking at pixels in a small region of the image. One such operation, for example, calculates the value of a pixel by averaging the values of the eight pixels that surround that pixel. When this is applied to each pixel in an image the result is a *blurring* or *smoothing* effect on the image. Figure 7-5(a) shows the lane close-up view from Figure 7-3. Figure 7-5(b) shows the results of a blurring operation performed on the lane close-up image.

Another useful spatial operation is one that sharpens the image (e.g., Figure 7-5(c)). In this case, the central pixel of the neighborhood is the sum of the differences between the central pixel's original value and the value each of its eight neighbor pixels.

A convenient way to express spatial operations of this sort is as a *convolution operation*. The convolution operation makes use of a *convolution kernel G*, a $q \times r$ matrix. Typically $q = r$, and $r \ll n$, $r \ll m$, where $n \times m$ are dimensions of the image I. No matter what the kernel contains, the convolution H is always defined for each point (x, y) in the image I as:

$$H(x, y) = \sum_{i=0}^{q-1} \sum_{j=0}^{r-1} I\left(x + i - \frac{r}{2}, y + j - \frac{q}{2}\right) G(i, j)$$

Figure 7-6 shows a sharpening kernel and a portion of an image. The kernel is applied to each pixel in succession starting at the top left and working to the bottom right. The pixel value at (x, y) is replaced by the sum of the product of the (i, j) element of the kernel with the $(x + i - r/2, y + j - 1/2)$ pixel of the image for all i and j.

Depending on the contents on the kernel, convolution can be used to transform images in many ways [5,18]. In the case of the lane following example we have been working on, we had remarked previously that the thresholded image held a lot of 'noise' in addition to the lane marking (See Figure 7-5). Smoothing the image with a 7×7 kernel generates a much less noisier image (See Figure 7-7) without losing any of lane marking. Let's consider what it would take to add this operation to our program.

Convolution kernel

-1	-1	-1
-1	8	-1
-1	-1	-1

Image

50	10	55	30	20
18	20	40	35	30
19	18	30	40	50
18	18	20	90	80
17	16	40	80	100

→

Kernel applied left to right, top to bottom

Figure 7-6: 3×3 Image convolution example.

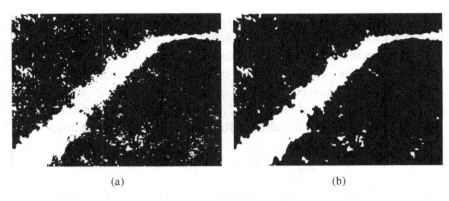

(a)   (b)

Figure 7-7:   (a) Original thresholded image and (b) smoothed image.

7.2.5.   *Partitioning*

The issue with carrying out a spatial operation in parallel is that each processor needs access to lines of the image that will have been scattered to other processors. To carry out a 3×3 smoothing on a row stripe of size n/p we need access to $n/p + 2$ lines of the image because we need the line before the first line in the strip and the line after the last line in the strip. There are two choices for partitioning:

(1) Scatter $n/p + 2$ overlapping row strips to each processor, but gather only an n/p row strip from each. The additional scatter communication cost for doing this is $2(3m)(p-1)/p\beta$.

(2) Scatter and gather n/p row strips as before, but then have each adjacent pair of processors do point-to-point communication of the overlapping strips. Each of the $p - 1$ processors does a send/receive and then a receive/send operation. All pairs do this in parallel so the cost is just the cost of two send operations, or $2(3m)/\beta$.

Since $(p - 1)/p$ is always less than one for $p > 1$, the time for the overlapping scatter approach is always smaller.

7.2.6.   *Program Design*

The overlapping scatter is implemented by setting up the displ and dendcnt arrays so that the row strip block size for each processor is two rows bigger than usual, and the starting location for the block is one row earlier than usual. (It is necessary to set first and last strips differently.)

```
for ( i = 0; i < p; i++ ) {
        if ( i==0 ) { offset = 0; incr=1; } /* first stripe */
        else if ( i==p-1 ) { offset =1; incr=1; } /*last strip */
        else { offset =1; incr=2; }  /* in between */
        displ[ i ]   = (BLOCK_LOW(id, p, n) - offset)*3*m;
        dendcnt[ i ] = (BLOCK_SIZE(id, p, n) + incr)*3*m;
}
```

The convolution operation is implemented using a for loop. We will assume that the convolution mask **G** is already present on all processors. The startOffset and endOffset values ensure that for the processors handling the first and last row strips, the convolution starts and ends with valid rows.

```
for ( y = startOffset; y < (BLOCK_SIZE(id, p, n)-endOffset); y++ ) /* row indx */
    for ( x = 1; x < 3*(m-1); x++ ) /* column index */ {
        sum = 0;
        for ( i = 0; i < 3; i++ ) for ( j = 0; j < 3; j++ ) {
            index=(y + j – 1)*3*m+(x + i - 1); /* 3x3 convolution, (1,1) is the center*/
            sum += image_l[index] * G[ i ][ j ];
        }
        result_l[ y*m+x ] = sum;
}
```

The displacements and end count arrays need to be recalculated as before to return the result image.

The next step in processing for a lane following application such as this might be to carry out a Hough filter on the thresholded image, to fit lines to the points identified as being on the lane markings. We have already discussed parallel approaches to the Hough transform for sonar data in Chapter 5, and the same approaches can be applied here. Once image lines have been detected, these can be transformed to the robot ground plane using a perspective projection to map the 2D image plane points onto the 2D ground plane. This perspective projection can be calculated in advance using a calibration pattern [5]. Parallel approaches to coordinate transformations were discussed in Chapter 4 and these approaches apply equally well to Hough line coordinate transformations.

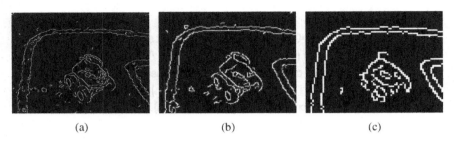

(a)   (b)   (c)

Figure 7-8:   Multiscale edge operations on the scene in Figure 7-3(a).

7.3.   Multiscale Image Processing

There is an advantage to inspecting an image at several scales, because the objects in an image can vary in apparent size due to visual perspective. It is useful to combine the results of applying pixel-based image operations at several different image scales. Consider the three images shown in Figure 7-8. These show the result of processing the scene in Figure 7-3(a) with a pixel-based edge operation, an operation that identifies the intensity changes that typically occur at the edges of objects or color regions in an image and which can also be done using convolution.

The identified edges show up as white pixels in these images. The operation was performed with the same kernel size on the original sized image (the leftmost image in Figure 7-8), then on an image with half the height and width dimensions of the original (the second image on the left), and so forth, from left to right. The result images have all been resized identically in Figure 7-8 for ease of comparison. The image on the left shows the most detailed breakdown of edges: details of the wheels of the robot are visible as is texture along the white lane lines. The image on the right the least detail: the robot is visible mainly as a rectangular region and the white line lines are devoid of internal detail.

A number of multiscale processing operations have been developed. *SIFT features* (Scale Invariant Image Transform features) [11, 22] use a multiscale measurement to identify image features that are unique across a range of scales, image rotations and translations. Ouerhani *et al.* [17] use a multiscale architecture to reliably identify salient objects in an image.

7.3.1. *Partitioning*

The serial implementation of a multiscale image operation would proceed by processing each scaled image in turn. Let there be s scales. The full sized image will be n rows by m columns. Each successive image will be half the row size and half the column size of the next larger image. If the time to process the full size image is $nm\chi$, then the next smaller scaled image will take $nm\chi/4$ and the next smaller will be $nm\chi/16$ and so forth. This is a geometric progression with ratio 1/4. The total time taken to process s scales will therefore be

$$T_{ser} = nm\chi + \frac{nm\chi}{4} + \frac{nm\chi}{16} + \cdots + \frac{nm\chi}{4^{s-1}} = nm\chi \left[\sum_{i=0}^{s-1} \frac{1}{4^i} \right]$$

$$= nm\chi \left[\frac{1 - \frac{1}{4^{s-1}}}{1 - \frac{1}{4}} \right] = nm\chi\xi(s)$$

Where $\xi(s)$ is the sum to s terms of the geometric progression. If we partition this processing on the cluster with row stripes as we did before, we have a parallel calculation time of $\frac{nm\chi\xi(s)}{p}$ since each processor just does the processing for every scale on its row strip of the image in parallel. The communication time for this approach is the same as for the previous single-scale, pixel-based operations.

However, it is also possible to partition this processing based on scale rather than based on row stripes. Let us assume that the s scale images are available for distribution to processors. We partition the available processors between the scales. Let P_i be the set of processors allocated to scale i. The root can distribute the ith scale image to one of the processors in P_i, which acts as a local root for that set of processors, and which will then scatter the image by row strip among all the processors in the set. Is this approach better than the simple row-strip approach for multiscale processing?

It is first necessary to establish a way to partition the processors among the scales. Since we have adopted a scale pyramid already it seems sensible to use the same criteria for allocating processors. We allocate the most processors for the full size image and the least for the smallest size image. We will use a geometric progression with ratio $1/4^i$ to partition the p

Table 7-1: Multiscale communication cost comparison.

	Row strip scattered	*Scale and row strip scattered*
Latency term	$\lambda \log p$	$\lambda \log s + \lambda \log \left(\dfrac{p}{\xi(s)} \right)$
Bandwidth term	$\dfrac{nm(p-1)}{p\beta}$	$\left[\dfrac{nm\xi(s)}{s} \right] \dfrac{(s-1)}{s\beta} + \dfrac{nm(p-\xi(s))}{p\beta}$

processors in the sets P_i for $i \in \{1, \ldots, s\}$:

$$|P_1| = \frac{p}{\xi(s)}, |P_2| = \frac{p}{4\xi(s)}, |P_3| = \frac{p}{8\xi(s)}, \ldots, |P_s| = \frac{p}{4^{s-1}\xi(s)}$$

Each partition will process its scale in parallel, so the calculation time in this case is the time for the largest scale (taking the longest time) to finish:

$$T_{cal} = \frac{nm\chi}{|P_1|} = \frac{nm\chi\xi(s)}{p}$$

This is the same as for the simpler row stripe partition, so there is no calculation advantage in partitioning by scale.

The communication cost of partitioning by scale will involve the cost for the root to scatter the scale images to each partition's root. Averaging the scatter size to $\frac{nm\xi(s)}{s}$, the scatter time is then

$$\lambda \log s + \left[\frac{nm\xi(s)}{s} \right] \frac{(s-1)}{s\beta}$$

Following this initial scatter, each partition will do a row stripe scatter. Again, we need only look at the one that takes the longest time. That is, the one with the largest scale, and that is done on the P_1 partition:

$$\lambda \log |P_1| + \frac{nm(|P_1|-1)}{|P_1|\beta} = \lambda \log \left(\frac{p}{\xi(s)} \right) + \frac{nm(p-\xi(s))}{p\beta}$$

Comparing these costs to the equivalent communication cost to scatter the full image to all processors, it's not difficult to see that the scale partitioning is always worse.

7.4. Video Tracking

Tracking is the process of monitoring and predicting the motion or location of a target object using sensory information (see [2] for a comprehensive overview). A robot may need to track significant aspects of its environment. For example, a robot operating with a team of other robots may need to track the locations of its team members. We will look at one approach to visual tracking. A robot looking at visual landmarks for place recognition (Chapter 6) for example may need to track the relative location of a landmark as the robot moves in the area around the landmark.

Let's say the robot is keeping track of targets t_1 through t_n up to some point in time. An image taken at the next point in time will yield some candidate target locations, c_1 through c_m. Data association [20] is the process of determining which candidates should be associated with which targets. Let us suppose we have a function $s(t, c)$ that captures how similar the candidate is to the target. In *Nearest Neighbor* data association, the correct data association for each target is chosen as

$$a_j = \max_i s(t_j, c_i)$$

Of course, the candidate that appears to be the best at one point in time might later turn out to be incorrect. Tracking algorithms such as the Joint Probability Data Association Filter (JPDAF) [19] and Multiple Hypothesis Tracking (MHT) [6] address this issue. Isard and Blake [10] introduced an approach to tracking called Condensation that is based on particle filters. A particle filter approach is particularly appropriate for parallel implementation as we discovered in Chapter 6 and in this section we will look at how we can implement Condensation in MPI. First, however, we need to discuss how we can implement the similarity function.

7.4.1. *Spatial Histograms*

A histogram is a measurement of the frequencies with which pixel values appear in an image. Lets us consider a gray level image, that is, an image with just one color channel[b]. Let $I : P \rightarrow V$ be a function that returns the value $v \in V$ of a pixel at a location $p \in P$ in the image. The *histogram of I* captures the number of times each pixel value occurs in the range of the function. Consider a set B of equivalence classes on V. A histogram of I, written h_I, maps B to the set $\{0, \dots, |P|\}$ such that

---

[b]As on a 'black and white' TV — which actually shows a gray level image, not a black and white (binary) image.

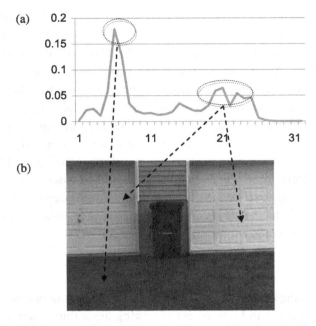

Figure 7-9:   (a) Image histogram (freq. vs. pixel value) (b) of gray-level image.

$h_I(b) = n_b$ and

$$n_b = \eta \sum_{i=1}^{|P|} \delta_{ib}$$

where δ_{ib} is equal to 1 *iff* the ith pixel is in the bth equivalence class and 0 otherwise, and η is a normalizing constant. The equivalence classes are called *bins* or *buckets*. Figure 7-9(a) shows a histogram of a gray-level scene, shown in Figure 7-9(b). The horizontal axis is the bin number. This histogram has 32 bins (so each bin covers a range of 8 consecutive pixel values). The vertical axis is the normalized frequency. The large peak is due to the dark ground in the bottom of the image in Figure 7-9(b). The two smaller peaks are the bright garage doors in the image.

A histogram for a region within an image will return the same information no matter the orientation or translation of that region in the image, and if the region size can be normalized, then the histogram will be invariant to scale as well. Because the location, rotation and scale of a landmark will change in an image due to perspective and robot motion, invariance to these changes is a good feature for a landmark tracking similarity measurement.

Unfortunately, a histogram is also invariant to the relative location of the pixel values in the region: A red and white checkerboard may give exactly the same histogram as a set of similarly sized red and white strips. One approach to minimizing the undesirable invariance properties of histograms is to use spatiograms. A *Spatiogram* extends the concept of a histogram by adding a normal distribution to each bin that gives the spatial concentration of the pixels that contribute to that bin:

$$h_I(b) = \langle n_b, \mu_b, \Sigma_b \rangle$$

where μ_b and Σ_b are the spatial mean and covariance of the values in the class b. One spatiogram h can be compared to another h' with the normalized comparison [12]:

$$\rho(h, h') = \sum_{b=1}^{|B|} \varphi_b \sqrt{n_b n'_b}$$

where the normalizing term φ_b is given as follows

$$\varphi_b = 2(2\pi)^{0.5} |\Sigma_b \Sigma'_b|^{0.25} N(\mu_b; \mu'_b, 2(\Sigma_b + \Sigma'_b))$$

Returning to tracking: We will assume that target state consists of a location in the image, and we will calculate the spatial histogram on a fixed size region of the image centered on the target location. The above definitions are for single channel, gray level spatiograms. An easy way to extend to color spatiograms is by defining single channel spatiograms for each of the color channels of the image.

7.4.2. *Condensation*

The implementation of the CONDENSATION [10] video tracking algorithm will look very similar to the implementation of the MCL localization discussed in Chapter 6, with one twist: We will need a motion model for the target. If x_t is the state of the target (its location and velocity in image coordinates, for example) then the condition distribution $p(x_t|x_{t-1})$

is our motion model. We also need an *observation model*. If we predict that the target has state x_t then, when we measure the image at the location given by the state, and get a measurement z_t, our observation model is the distribution $p(z_t|x_t)$. If $h(x_t)$ is the spatiogram presenting the target and z_t is the spatiogram of the candidate region, then we can say

$$p(z_t|x_t) = \rho(z_t, h(x_t))$$

An advantage of Condensation over other tracking methods is that it places little constraints on the motion and observation model except that they can be evaluated on each particle. This advantage is bought at the cost of needing a large number of particles for accurate results. This is also the reason it is a good choice for cluster implementation since, as we have seen in Chapter 6, the operations on each particle are independent.

(1) The algorithm starts at step $t-1$ with a collection S_{t-1} of n particles x_i for $i \in \{1, \ldots, N\}$ representing the location of a target being tracked in the image.

(2) The *motion model* for the target: $p(x_i'|x_i)$ is used to propagate each particle's state. The new collection S_t' represents all the locations that the target could have moved to when starting from S_{t-1}.

(3) The observation model $p(z_i'|x_i')$ is used to weight the particle. The weights, w_i for $i \in \{1, \ldots, N\}$, are normalized, and a cumulative weight is calculated for each particle.

(4) The set S_t' is resampled so that higher weighted particles are more likely to be selected to produce S_{t+1}.

(5) The state of the target is calculated as the linear sum $\sum_1^N w_i x_i$.

7.4.3. *Partitioning*

We can partition the condensation algorithm in a fashion similar to our scheme for MCL Localization. We broadcast the $n \times m$ image to all p processors, and then partition the N samples equally among the p processors. The communication cost is $(\lambda + nm/\beta)\log p$ for the broadcast, and $\lambda \log p + N(p-1)/p\beta$ for the particle scatter.

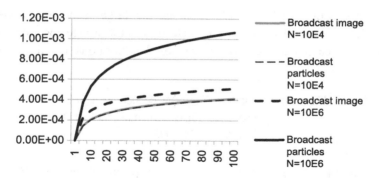

Figure 7-10: Performance comparison of condensation partitioning approaches.

Each processor then propagates its N/p particles with the motion model. It then calculates N/p spatiograms on fixed size regions around the locations of each particle. Finally it carries out N/p spatiogram comparison operations between the spatiogram.

An alternative partition scheme is to partition the image by row strips, and broadcast all n particles. Each processor calculates only the portion of propagations, spatiograms and comparisons that occur in its image strip.

Figure 7-10 shows that for smaller collections of particles ($N = 10^4$), for a fixed size (640×480) image, broadcasting particles and scattering the image has marginally smaller cost. However, for larger collections of particles ($N = 10^6$), broadcasting the image and scattering the particles has the better performance. We would prefer to work with larger collections of particles, since condensation produces a more accurate tracking in that case. But furthermore, scattering the image has the disadvantage that all the particles may end up in a small number of row stripes, leaving most processors idle.

7.4.4. *Program Design*

The coordination of the parallelism will be carried out by the root processor using the improved serial fraction method developed for Monte-Carlo Localization in Chapter 6. The root begins by initializing the set of particles to the initial target location and scattering the N particles in blocks to the p processors. Each particle includes the target state and a weight value. If, for example, the particle state is just the image location of the target, then each particle will consist of 3 contiguous locations in the array.

```
MPI_Scatterv( &samples[0],   displ, dendcnt, MPI_FLOAT,
              &samples_l[0], dendcnt[rank], MPI_FLOAT,
              0, MPI_COMM_WORLD );
```

The **samples** array is the entire list of particles, and **samples_l** is the processor's allocation of particles. The arrays **displ** and **dedcnt** are calculated using the block partition macros (based on N and not the size of the array, as that might split particles over processors). Note: if it is possible for each processor to independently initialize its block of particles, then this initial scatter is not needed. It is also fair to assume that the target spatiogram has already been broadcast to all processors as part of the tracking initiation.

The full image (for each image in the tracking sequence) needs to be present on all processors for the observation model step of processing.

```
MPI_Bcast ( image, n*m, MPI_BYTE, 0, MPI_COMM_WORLD );
```

Each processor must now traverse its block of particles. For each particle it carries out the motion model with **motionModel(samples_l[s])**. This routine predicts for each particle where it will have moved in this time step. We can assume, for example, that the target will move an arbitrary but small distance in an arbitrary direction from its previous position. If more information is available about the target dynamics, for example, that it will continue to move with constant velocity, then this information should be incorporated in **motionModel**.

The observation model processing involves two steps: First, the set of spatiograms corresponding to particles in **samples_l** are calculated. Second, the spatiogram comparison function is used to compare each particle's spatiogram with the target spatiogram, weighting the particle with the resulting value. Finally the block of particles assigned to the processor is sorted and the full list of particles gathered together onto the root processor.

```
for ( s = 0; s < dendcnt[rank]; s++ ) { /* traverse samples */
    motionModel( samples_l[s] );
    calculateSpatiograms( samples_l[s],
                              spatiograms[s], image );
    weightSample( samples_l[s], spatiograms[s],
                              targetSpatiogram );
    sort(samples_l);
}
MPI_Gatherv( &samples_l[0], dendcnt[rank], MPI_FLOAT,
             &samples[0], dendcnt, displ, MPI_FLOAT,
             0, MPI_COMM_WORLD );
```

The resampling step can be done as we did for MCL localization in Chapter 6. First the p sorted particle lists are merged to a single sorted list according to their weight. Then a cumulative value is calculated for each particle, and finally the resulting list is rebroadcast to all processors.

On each processor, blocks of new samples are selected by generating a random number between 0 and the sum of all weights. The particle with closest higher cumulative weight is then found with binary search and added as a new sample. In the following code, we will assume that each particle is represented by numP contiguous components, and that the offset of the weight component is offW.

```
if ( rank == root ) { /* only the root does this */
    merge(samples, displ, dendcnt);
    for ( s = 1; s < N; s++ ) { /* make cumulative score */
        samples[s*numP+offW]  = samples [(s-1)*numP+offW]
                                     +samples[s*numP+offW];
    }
}
MPI_Bcast( samples, N*numP, MPI_FLOAT, 0, MPI_COMM_WORLD);
resample(samples, displ, dendcnt); /* sublist resample */
```

7.5. Summary

Vision is an important sensory channel for robots but it is one that is surprisingly difficult to use effectively. The ease with which visual interpretation comes to humans belies the difficulty of doing the same thing on a robot. By looking at the pin-hole camera model, one source of difficulty becomes clear: the 2D projected image has lost important structural information about the 3D scene. By looking at how images are represented as a matrix of light intensities a second difficulty becomes clear: identifying and separating parts of a scene becomes quite difficult at the pixel level.

Although pixel-based computer vision algorithms are more typically done on fine-grained, shared memory machines [4], the growing prevalence of multiprocessor, multi-core machines prompts us to look at cluster implementations of some common iconic image processing operations. As an example, we looked at the problem of programming a robot to follow white lane lines on grass. We looked at the identification of the white lane lines in a visual image, and we developed a row stripe partition approach to extract the lane information in the image. Because all the pixel operations are independent this algorithm has good parallel time characteristics.

We also looked at another important class of iconic image process, operations that use a spatial neighborhood around a pixel. A convolution operation for smoothing and sharpening was the example operation developed in more detail. By scattering overlapping data, the independent nature of the processing, and hence the good parallel timing, was preserved at the cost of slightly higher communication. There are many, many image processing operations that share the characteristics of these two problems and can be easily extrapolated from them.

We also looked at multiscale iconic image processing which allows the identification of features at a variety of scales in the image. We compared a simple row strip partition with a more sophisticated scale and row strip partition and found no advantage to the latter.

In addition to individual vision algorithms developed using MPI, there are also special libraries for MPI processing of images. Examples include the SPMD image processing library [16] and PIPT [23]. Both of these provide the user with a higher level algorithm development environment than MPI does but restrict the problem solving paradigm. SPMD focuses on splitting parallel loops with little communication. PIPT provides a worker/manager framework in which chunks of the image can be processed in parallel by worker processors. Both provide ways to handle the issues of load balancing, to ensure processors are not left idle.

Tracking is a widely studied field in robotics and computer vision and there are many approaches in the literature. The particle filter is a popular approach because of its ease of implementation and lack of constraints on the target motion model. We looked at using a spatial histogram target model and partitioning the particle collection among processors (as we did in Chapter 6 for MCL localization). Even though particle filtering has a high computational load, it is straightforward to implement this in parallel on a cluster. Medeiros *et al.* [14] also report a parallel SIMD implementation of particle filter color tracking on a smart camera.

References

1. Arbib, M.A., *The Metaphorical Brain 2*, Wiley & Sons 1989.
2. Bar-Shalom, Y. and Fortmann, T., *Tracking and Data Association*. Academic Press 1988.
3. Birchfield, S.T. and Sriram Rangarajan, Spatial Histograms for Region-Based Tracking, *ETRI Journal*, V29, N5, October 2007.
4. Braunl, T., Feyrer, S., Rapf, W., and Reinhardt, M., *Parallel Image Processing*, Springer-Verlag, 2001.
5. Bradski, G. and Kaehler, A., *Learning OpenCV*, O'Reilly 2008.
6. Cox, I.J. and Hingorani, S.L., An Efficient Implementation and Evaluation of Reid's Multiple Hypothesis Tracking Algorithm for Visual Tracking. *Int. Conf. on Pattern Recog.* (1994) pp. 437–442.
7. Dudek, G. and Jenkin, M., *Computational Principles of Mobile Robotics*, Cambridge Press 2000.
8. Klechenov, A., Gupta, A.K., Wong, W.F., Ng, T.K., and Leow, W.K., Real-time Mosaic for Multi-Camera Videoconferencing, *MIT-Singapore Alliance Symposium*, Singapore 2003.
9. Kuhn, M., Parallel Image Registration in Distributed Memory Environments, *M.S. Thesis*, Swiss Federal Institute of Technology, 2004.
10. Isard, M. and Blake, A., Condensation — Conditional density propagation for visual tracking. *IJCV: International Journal of Computer Vision* 29 5–28, 1998.
11. David G. Lowe, Distinctive image features from scale-invariant keypoints, *International Journal of Computer Vision*, 60(2), pp. 91–110, 2004.
12. Lyons, D.M., Sharing Landmark Information using Mixture of Gaussian Terrain Spatiograms, *IEEE/RSJ International Conference on Intelligent RObots and Systems (IROS)*, St. Louis, MO, October 2009.
13. Lyons, D. and Hsu, D.F., Rank-based Multisensory Fusion in Multitarget Video Tracking. *IEEE Intr. Conf. on Advanced Video & Signal-Based Surveillance*, Como, Italy, 2005.
14. Medeiros, H., Gao, X.T., Kleihorst, R.P., Park, J., and Kak, A.C., A parallel color-based particle filter for object tracking, *The 6th ACM Conference on Embedded Networked Sensor Systems*, Raleigh NC, November 5–November 7, 2008.

15. Montemerlo, M. *et al.*, Junior: The Stanford Entry in the Urban Challenge, *Journal of Field Robotics* Volume 25, Issue 9 (September 2008).
16. Oliveira, P. and du Buf, H., SPMD Image Processing on Beowulf Clusters: Directives and Libraries, *Proceedings of the 17th International Symposium on Parallel and Distributed Processing*, 2003.
17. Ouerhani, N., von Wartburg, R., Hugli, H., and Muri, R., Empirical Validation of the Saliency-based Model of Visual Attention, *Electronic Letters on Computer Vision and Image Analysis* 3(1):13–24, 2004.
18. Parker, J., *Algorithms for Image Processing and Computer Vision*, Wiley 1997.
19. Rasmussen, C. and Hager, G., Joint Probabilistic Techniques for Tracking Multi-Part Objects. *Proc. Computer Vision & Pattern Recognition. Santa Barbara*, CA; (1998) pp. 16–21.
20. Rao, B., *Data Association Methods for Tracking Systems*, in *Active Vision*, Blake, A., Yuille, A., Editor. 1992, MIT Press. pp. 91–106.
21. Rosselot, D. and Hall, E., Processing real-time stereo video for an autonomous robot using disparity maps and sensor fusion, *SPIE Intelligent Robots and Computer Vision XXII: Algorithms, Techniques, and Active Vision*. pp. 70–78, 2004.
22. Se, Stephen, David Lowe and Jim Little, Local and Global Localization for Mobile Robots using Visual Landmarks, *Proceedings of the International Conference on Intelligent Robots and Systems* (IROS), pp. 414–420, 2001.
23. Squyres, J., Lumsdaine, A., and Stevenson, R., A toolkit for parallel image processing, *Proceedings of the SPIE Conference on Parallel and Distributed Methods for Image Processing*, 1998.
24. Wilkinson, B. and Alen, M., *Parallel Programming*. 2nd Ed., Prentice-Hall 2005.

Chapter 8

Learning Landmarks

We have previously discussed the issues involved in building metric, occupancy grid maps (Chapter 6). Recall that one of the key purposes for having the robot learn a map is to allow it to localize (determine where it is). A second key purpose is to carry out better motion planning by using knowledge about the space in which the motion will occur. However, an issue with occupancy maps is that their storage requirements scale with the square of the area to be represented. Even worse, much of this area may not be interesting or relevant for the tasks that the robot must carry out — it may simply be area through which the robot must pass to get to destination.

Consider your morning commute. Unless you are lucky enough to work at home, you probably have to drive on some local roads from your house until you get to a larger road or highway. From this, you sooner or later take an exit and proceed on some local roads to your place of work. It is important to have good map information whenever you have to make a choice along this route: for example, switching between local roads, or to get on, or off, the highway. However, the details of the 'choiceless' sections of your commute are not, in general, particularly important. In fact, you may not even be able to recall these details at will unless they strike you for some other reason.

In Chapter 6 we also introduced the idea of a *topological map*: a graph-like collection of *places*. Each place is identified by sensory feature measurements (the place *signature*) and the places are linked together by routes without any absolute spatial measurements. Your recollection of your daily commute may be a bit more like this topological map than like an occupancy-grid map, except for those places where you need to make decisions. A reasonable approach to trade off the map storage requirements with map accuracy requirements is as follows: Use a local metric map representation *within* places, but recognize places with sensory signatures, and store just the route connectivity information *between* places. The route information allows the robot to navigate from one place to another. However motor skills or other behaviors (Chapter 9) may be necessary to successfully navigate the transition from one place to another.

Consider the landscape shown in Figure 8-1(a): roads traverse the scene, but at places where road intersections occur, visual landmarks are noted. These visual landmarks are the signatures for the places at which decisions need to be made. An equivalent topological map is shown in Figure 8-1(b). The places are labeled with the landmark number. The details of the routes between places are not represented, just the fact that one place is reachable from another. The map can store more detailed metric information for each place. For example, the map would have detailed information about the major intersection for the place labeled 4 in Figure 8-1.

In this chapter we will investigate some approaches to learning land-marks to identify places. Although any sensory signature could be involved in the identification of a place, we will restrict ourselves to looking at visual signatures. Our objective will be to take multiple measurements of each of our landmarks and build a model (the signature) for each landmark. When an image of an unknown landmark is presented to the robot, we would like it to use its collection of landmark models to determine to which, if any, of the landmarks the image corresponds. This problem falls into the

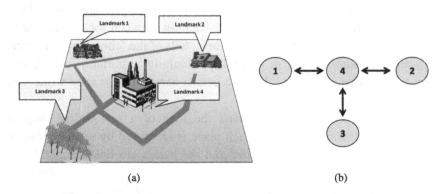

Figure 8-1: (a) Scene with landmarks, and (b) equivalent topological map.

class of *machine learning* problems and is an example of what is called *unsupervised learning*. Clustering (not to be confused with computational clusters) is an approach to solving unsupervised learning problems of this kind. The clusters refer to subsets of the collection of data points in which all the members are 'close' to one another in some measurement space. We will look at two fairly common approaches to clustering: *K-Means* and *Expectation-Maximization (EM)*.

8.1. Landmark Spatiograms

Spatiograms were introduced in Chapter 7 to represent the visual similarity between a target candidate and a target. We can also employ spatiograms to recognize landmarks [8–10]. Consider Figure 8-2: Figure 8-2 column (a) shows images of three landmarks: a garbage bin, a wishing-well

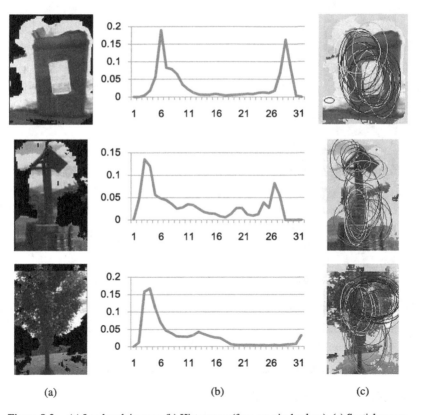

(a)   (b)   (c)

Figure 8-2: (a) Landmark images, (b) Histograms (freq. vs. pixel values), (c) Spatial means.

and a tree. The images were taken with a stereocamera[a] and the image foreground region identified by depth filtering. The region is shown in the figures by the irregular image boundary on a black background. Spatiograms were calculated for each foreground region in Figure 8-2 column (a), and the normalized frequency information displayed as a histogram for each row in Figure 8-2 column (b). The spatial mean and variance is displayed in Figure 8-2 column (c) as an ellipse with the spatial mean as center and (assumed independent) x and y variances as major and minor axis. The brighter the contour of the ellipse, the higher the frequency value associated with that ellipse.

Figure 8-3 shows three spatiograms of the same landmark from slightly different views. There is a visible similarity between the shape of the three

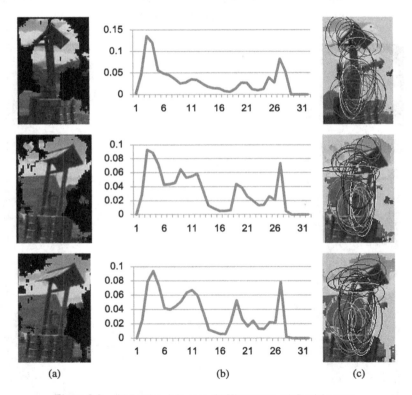

(a)   (b)   (c)

Figure 8-3:   (a) Landmark images, (b) Histograms, (c) Spatial means.

---

[a] Videre Design model STH-MDCS2 on a TracLabs Biclops PT base.

histograms, and also between the spatial areas that produce the highest frequency information (the white ellipses around the bucket and the well posts in Figure 8-3 column (c)). Our approach to learning landmarks will be to take many examples of each landmark and to try to group these examples into clusters that are close together based on the spatiogram similarity measure ρ. When we are given candidate landmarks, we will be able to identify which landmark it is by determining into which cluster the spatiogram of the candidate falls.

8.2. K-Means Clustering

K-Means is an algorithm for partitioning data points into clusters in such a way that the sum of the distances between points in the same cluster is minimized. If there are n points, p_i for $i \in \{1, \dots, n\}$, and there are m clusters, c_j for $j \in \{1, \dots, m\}$ then the K-Means algorithm iterates two passes:

- Pass 1: Assign each p_i to a cluster c based on

$$c = \min_j \text{dist}(p_i, c_j).$$

- Pass 2: Recalculate each c_j cluster center as

$$\sum_{i \in c_j} \frac{p_i}{|c_j|}$$

The iteration can stop when the cluster centers in Pass 2 no longer change. If each point is actually a spatiogram, then the distance measure that we need to use is ρ and the recalculation of cluster centers is an averaging of spatiograms. A good selection of initial cluster center locations is typically important. One approach is to select points that have the largest sum of distances from all other points as the initial cluster center locations.

Recall that we define a spatiogram as $h_1(b) = \langle n_b, \mu_b, \sum_b \rangle$. We can now define the average of two spatiograms h and h' as

$$av(h, h') = \left\langle \frac{n_b + n'_b}{2}, \frac{\mu_b + \mu'_b}{2}, \frac{\sum_b + \sum'_b}{4} \right\rangle$$

Since the sum of n independent normal distributions is a normal distribution, we can generalize this average to carry out pass two of the K-Means algorithm. (For a 32-bin single color-channel spatiogram with independent x and y spatial distributions, this sum involves 5×32 additions.) The cluster centers are spatiograms that are 'prototypes' for the landmarks that they represent.

8.2.1. *Partitioning*

The cluster to point distances are all calculated independently in K-Means. There is therefore the opportunity to carry out this phase in parallel [12]. Typically $m \ll n$, that is, there will be much fewer clusters than points. In terms of landmark clustering: we will have many example images (image spatiograms) of a small number of landmarks (cluster spatiograms). Hence, we should broadcast the cluster spatiograms and scatter the image spatiograms. The image spatiograms only have to be scattered *once* (before the first iteration); the cluster spatiograms need to be rebroadcast on every iteration (since the cluster centers may move on every iteration).

Each processor will calculate the m distances for its n/p image spatiograms and record the cluster chosen for each image spatiogram — concluding the first pass. Each processor will then sum all the image histograms for each cluster and participate in a reduction by sum on each cluster spatiogram. The root completes the averaging for all clusters and rebroadcasts the results.

8.2.2. *Program Design*

Let us consider the data points to be stored as an array of n spatiograms, points. Similarly, the cluster centers will be an array clusters of m spatiograms. The initial scatter of data points is accomplished with

```
MPI_Scatterv(   &points[0],  displ, dendcnt, MPI_FLOAT,
                &points_l[0], dendcnt[rank], MPI_FLOAT,
                0, MPI_COMM_WORLD );
```

The values of displ and dendcnt arrays are calculated using the block macros to divide n over the p processors in a balanced fashion. This will scatter (approximately) sn/p floating point values to each processor, where s is the size of the spatiogram structure. We will assume a procedure initClusters(clusters, points) that establishes the initial cluster centers distributed evenly among the data points. The main loop of the clustering will iterate through broadcasting cluster centers, partitioning points to clusters, and reducing the local clusters:

```
initClusters( newClustersG, points );
/* set up clusters spread among data points */
zeroClusters( clusters );
while ( changed( clusters, newClustersG) ) { /* if the two are different, cont. */
        copyClusters( newClustersG, clusters ); /* set up clusters */
        MPI_Bcast ( clusters, m*s, MPI_FLOAT, 0, MPI_COMM_WORLD );
        zero( clCounts );
        zero( clMembers );
        assignPoints( points_l, clusters, clMembers, clCounts );
        averageClusters( points_l, clMembers, clCounts, newClusters );
        /* reduce the local clusters sums to global sums,
            also reduce the count in each cluster */
        MPI_Reduce( newClusters, newClustersG, m*s, MPI_FLOAT,
                            MPI_SUM, root, MPI_COMM_WORLD);
        MPI_Reduce( clCounts, clCountsG, m,
                            MPI_INT, MPI_SUM, root, MPI_COMM_WORLD);
        If ( rank == root )  /* finish the averaging of clusters */
            for ( j = 0; j < m; j++ )
                averageSpatiogram( newClustersG[ j ], clCountsG[ j ] );
    }
```

The **assignPoints** procedure steps through the points allocated to a processor and finds the closest cluster center for each point.

```
for ( i = 0; i < BLOCK_SIZE( rank, p, n); i++ ) {
        /* all points assigned to this processor*/
        for ( j = 0; j < m; j++ ) /* each cluster */
                dist[ j ] = rho( point_l[ i ], cluster[ j ] );
                /* spatiogram comparison */
        c = minindex( dist ); /* index of mind value */
        clMember[ clCounts[c]++ ] = i; /* record point for cluster c */
    }
```

The procedure **averageClusters** steps through all the points allocated to each cluster on each processor and averages them.

```
zeroClusters(newClusters);
for ( j = 0; j < m; j++ ) /* for each cluster */
    for ( i = 0; i < clCounts[ j ]; i++ ) /* for all pts assigned to this cluster*/
        addSpatiogram( newCluster[ j ], points_l[ clMember[ i ] ] );
```

8.2.3. *Analysis*

The serial implementation of K-Means carries out the averagePoints and averageClusters routines for all the data points on each iteration. If the time to carry out the spatiogram comparison is χ_1 then assigning points takes time $nm\chi_1$. If the time to add spatiograms is χ_2 then the time to average the clusters is $m(n/m)\chi_2$ assuming an average of n/m points per cluster. The serial time is given by $T_{ser} = n(m\chi_1 + \chi_2)$.

In the parallel implementation, the same calculations need to be performed for n/p points (simplifying to the case where p divides n evenly). The calculation time is $T_{cal} = n(m\chi_1 + \chi_2)/p$. The scattering is just done once, when K-Means initializes, so we will not count its contribution to the parallel time. However, in each iteration, the broadcast and the reduce operations need to be performed for all clusters. The communication time is $T_{com} = 2(\lambda + ms/\beta)\log p$. Hence, the total parallel time is

$$T_{par} = \frac{n(m\chi_1 + \chi_1)}{p} + 2\left(\lambda + \frac{ms}{\beta}\right)\log p$$

Let the serial time be c and the parallel communication time be k. In that case, the speedup can be written as

$$\psi = \frac{T_{ser}}{T_{par}} = \frac{c}{\frac{c}{p} + k} = \frac{1}{\frac{1}{p} + \frac{k}{c}}$$

If the ratio $k/c > 0$ then the speedup is less than linear $\psi < p$; the smaller the k/c ratio the closer the speedup is to linear. Looking at the k/c ratio for this problem, we have

$$\frac{2\left(\lambda + \frac{ms}{\beta}\right)\log p}{n(m\chi_1 + \chi_1)} \cong \frac{\frac{2ms}{\beta}}{nm\chi}\log p = \frac{2s}{n\beta\chi}\log p$$

Surprisingly the number of clusters is not as significant to the speedup as the number of data points. This ratio tells us that speedup will improve as the number of points increases.

8.3. EM Clustering

A second common algorithm used to assign points to clusters is *Expectation-Maximization* or *EM* [5]. EM tries to find a mixture of Gaussian functions that cluster the data points. It differs from K-Means in that it doesn't make a 'hard' assignment of each data point to a cluster. Instead, data points can have degrees of membership in every cluster. Since each cluster is a Gaussian, it is characterized not just by a center (the mean) by also by a variance. For a 'gentle' introduction to EM see [3].

EM can also be described as a two-pass iterative method and it is quite similar to K-Means in structure. The goal of EM is to find the *maximum likelihood* solution for clustering the data points. The algorithm iterates through an *Expectation* step, where the data point associations with the Gaussian clusters are calculated, and a *Maximization* step, where the cluster Gaussians are redefined based on the data point associations.

The probability of a point belonging to a cluster is described as a mixture of Gaussians $N(\mu_j, \sum_j)$ for $j \in \{1, \ldots, m\}$:

$$p(x) = \sum_{j=0}^{m} w_j N\left(x; \mu_j, \Sigma_j\right)$$

In the Expectation step, the matrix of probabilities is calculated:

$$Q = [N\left(x_i; \mu_j, \Sigma_j\right)]_{i=1,\ldots,n,\ j=1,\ldots,m}$$

From this table we can calculate matrix R, where:

$$R = [r_j(x_j)]_{i=1,\ldots,n\ j=1,\ldots,m}$$
$$r_j(x) = \frac{w_j N(x; \mu_j, \sum_j)}{\sum_{k=1}^{m} w_k N(x; \mu_k, \sum_k)}$$

In the Maximization step, the means μ_j, variances $\sum_j$ and weights w_j are recalculated:

$$w_j = \frac{1}{n} \sum_{i=1}^{n} r_j(x_j)$$

$$\mu_j = \frac{\sum_{i=1}^{n} x_i r_i(x_i)}{\sum_{i=1}^{n} r_i(x_i)}$$

$$\sum_j = \frac{\sum_{i=1}^{n} r_j(x_i)(x_i - \mu_j)(x_i - \mu_j)^T}{\sum_{i=1}^{n} r_i(x_i)}$$

We can apply this algorithm to cluster histograms (but not spatiograms) by considering a histogram as a vector and $N(\mu_j, \sum_j)$ as a Gaussian function of that vector.

8.3.1. *Partitioning*

The expectation maximization phase of EM, generating Q, can be partitioned among processors by broadcasting the cluster information for the m clusters, and scattering the n data points among the p processors. We expect again that $m \ll n$. Each processor can generate all the cluster probabilities for each data point completely independently and the $r_j(x)$ values just for its share of the data points.

In the maximization phase, the local sum of $r_j(x)$ and $xr_j(x)$ values can be subtotaled locally on each processor. They are then reduced to the root to get the new weight and mean values. Finally, the local calculations for $r_j(x)(x - \mu)(x - \mu)^T$ are completed and reduced to the root.

If m is large enough ($m > p$ at least) then it can make sense to partition the clusters among processors for the maximization step [7].

8.3.2. *Program Design*

The data points are scattered to processors before the first iteration. The means, weights and variances for each cluster are (re)broadcast on every iteration. The expectation step can proceed in parallel without communication. Each processor evaluates all the cluster probabilities for each of the points it has: i.e., it evaluates a block of rows of the matrix Q. The same block of rows of the matrix R can then be evaluated.

The maximization step requires that processors share intermediate results. Let the variable rSum_l [j] contain the R matrix column partial sum for the jth column (jth cluster) for the points allocated to a processor. The global sum is calculated as:

```
MPI_Reduce( rSum_l, rSum, m*s, MPI_FLOAT,
            MPI_SUM, root, MPI_COMM_WORLD);
```

The root can then calculate the new cluster weights by dividing by the number of points n. Similarly, to calculate the new means, the partial sums $x_i r_j(x_i)$ need to be reduced. The global sum rXSum is calculated from the local sums rXSum_l[j] in the same fashion as rSum. Since the root already has rSum, it can calculate the new cluster means.

The variance calculation requires that the means be available to all processors. One approach is to broadcast the means and weights after they are calculated and then proceed to the variance calculation. We will take a slightly different approach that allows us to introduce the MPI_Allreduce MPI function:

```
int MPI_Allreduce (    void          *sendbuf,
                       void          *recvbuf,
                       int           count,
                       MPI_Datatype  datatype,
                       MPI_Op        op,
                       MPI_Comm      comm ) .
```

The MPI_Allreduce routine has the same parameters as MPI_Reduce; however, its effect is to reduce the value(s) in sendbuff to the global reduced value in recbuff on *all* processors. If we use this routine for the weight and the mean reductions, then all processors can calculate the new weights and means in parallel.

With the means in place, each processor can calculate a partial sum of $r_j(x_j)(x_i - \mu_j)(x_i - \mu_j)^T$ and this can be reduced (*allreduced*) to calculate the new variances. Iteration proceeds until the Gaussian parameters do not change significantly.

8.3.3. *Analysis*

The time performance of EM has a similar form to that of K-Means. The serial time is given by:

$$T_{ser} = nm\chi_1 + nm\chi_2$$

where $nm\chi_1$ is the time to calculate the Q matrix while $mn\chi_2$ is the time to sum the R matrix columns and calculate the new weights, means and variances.

The expectation step has no communication needs. Hence, this can be partitioned among the cluster processors with (in the case p divides n evenly) $n\chi_1/p$ time. Part of the maximization can be done in parallel: the partial sums $r_j(x_i), x_i r_j(x_i)$ and $r_j(x_i)(x_i - \mu_j)(x_i - \mu_j)^T$ for a time $mn\chi_2/p$. However, three *allreductions* need to be performed to complete the maximization step, so the total time is (where $\chi = \chi_1 + \chi_2$):

$$T_{par} = \frac{nm\chi}{p} + 3 \times 2 \left(\lambda + \frac{ms}{\beta} \right) \log p$$

Looking at the k/c ratio, we see the same interesting result as for K-Means: the number of clusters has less effect on the speedup that the number of points.

The use of MPI_Allreduce does not alter the time of the program [4]: the $\log p$ time to reduce the value is followed by a second $\log p$ time to broadcast it again.[b] However, the processor utilization does improve, since each processor calculates the weights and means in parallel rather than waiting for the root to calculate and then broadcast these values.

8.4. Summary

If a robot navigates over a very large area, it may not be possible for it to store detailed information about all spatial locations. An alternate approach is to store detailed information only on places where the robot needs to be able to make navigation decisions, but store little or no information on other locations. These detailed descriptions can be represented as the *places* in a topological map. However, the robot then needs to be able to

---

[b]A more efficient implementation of Allreduce (Allgather and Allscatter) depends on many factors including network connectivity [13].

recognize when it has arrived at a place. In this chapter, we looked at an approach to this issue where we characterized places by visual landmarks. Each landmark was represented by an image spatiogram. In the last chapter, we introduced the spatiogram comparison operation, so the problem in this chapter becomes how to learn landmark spatiograms.

We approached this problem using an unsupervised machine learning approach called clustering (to be distinguished from computational clusters). A collection of many images of each landmark are taken, and a clustering algorithm is used to characterize how these images group together in terms of the similarity between their spatiograms. These 'groups' allow us to build landmark recognition models.

We looked at the problem of learning visual landmarks using parallel implementations of two unsupervised machine learning algorithms, K-Means and EM (Expectation-Maximization) using a mixture of Gaussians. K-Means is a simpler and faster algorithm, but because it makes 'hard' assignments of data points to clusters, the quality of the clustering may not be as good as EM. In both cases, an analysis of the speedup indicates that the number of clusters does not play much of a role in speedup, whereas speedup improves for larger numbers of data points.

References

1. Arkin, R.C., *Behavior-Based Robotics*, MIT Press 1998.
2. Bekey, G.A., *Autonomous Robots*, MIT Press 2005.
3. Bilmes, J., A Gentle Tutorial on the EM Algorithm Including Gaussian Mixtures and Baum-Welch, *ICSI Technical Report TR-97-021*, May 1997.
4. Bongo, L.A., Anshus, O.J., and Bjørndalen, J.M., Evaluating the Performance of the Allreduce Collective Operation on Clusters: Approach and Results, *Technical Report 2004-48*, University of Tromso, 2004.
5. Dempster, P., Laird, N.M., and Rubin, D.B., Maximum Likelihood from Incomplete Data via the EM Algorithm, *Journal of Royal Statistics Society*, vol. B-39, 1977.
6. Dudek, G. and Jenkin, M., *Computational Principles of Mobile Robotics*, Cambridge Press 2000.
7. López de Teruel P.E., García, J.M., and Acacio, M.E., The Parallel EM Algorithm and its Applications in Computer Vision. *Proceedings Parallel and Distributed Processing Techniques and Applications*, Las Vegas, Nevada 1999.
8. Lyons, D.M., Sharing Landmark Information using Mixture of Gaussian Terrain Spatiograms, *IEEE/RSJ International Conference on Intelligent RObots and Systems (IROS)*, St. Louis MO, October 2009

9. Lyons, D.M., Detection and Filtering of Landmark Occlusions using Terrain Spatiograms. *IEEE Int. Conference on Robotics and Automation*, Anchorage, Alaska, May 2010.
10. Lyons, D., Selection and Recognition of Landmarks Using Terrain Spatiograms, *IEEE/RSJ International Conference on Intelligent RObots and Systems (IROS)*, Tapei, Taiwan, October 2010.
11. de Souza, P.S.L., Sabourin, R., de Souza, S.R.S., and Borges, D.L., A low-cost parallel K-Means VQ algorithm using cluster computing, *Seventh International Conference on Document Analysis and Recognition, (ICDAR'03)*, Edinburgh Scotland, 2003.
12. Joshi, M.N., Parallel K-Means Algorithm on Distributed Memory Multiprocessors, *Computer*, April, 2003.
13. Patarasuk, P. and Yuan, X., Bandwidth optimal all-reduce algorithms for clusters of workstations, *Journal of Parallel and Distributed Computing* 69(2) (February 2009).
14. Thrun. S., *Robotic mapping: A survey*. In Lakemeyer, G. and Nebel, B. eds., *Exploring Artificial Intelligence in the New Millennium*. Morgan Kaufmann, 2002.

Chapter 9

Robot Architectures

Up to this point, we have looked at algorithms that address some key robot computational problems in motion, sensing and navigation, and we have analyzed their implementation and performance. This chapter switches perspective from algorithm to *architecture*. Architecture is most commonly used in Computer Science to describe the hardware of a computer system, usually in terms of functional components and their connectivity. The term is also used to describe the breakdown of a software system in functional components and connectivity. A *robot architecture* is typically a software architecture [5]; Arkin defines robot architectures as "software systems and specifications that provide languages and tools." [4]. He focuses his definition of architecture on *behavior-based systems* — a robust approach to constructing robot programs where the program is built as a collection of mostly independent, communicating modules responsible for physical robot behaviors. In this chapter, we will introduce several robot architectures for behavior-based systems and develop cluster implementations for them. This will require going beyond the Single Program Multiple Data (SPMD) program model that we have adopted so far.

9.1. Behavior-Based Robotics

The objective of robot programming is to build the software to control a robot to carry out some physical tasks, usually with the goal of achieving a desired physical state or sequence of states. For example, robots are used in industry for many jobs, including:

- to move parts of a product into place during manufacturing;
- to pack pallets of finished products for shipping;
- to unpack pallets of materials that will be used in manufacturing;
- to spray paint parts;
- to weld seams in subassemblies.

In applications other than manufacturing, mobile robots are used to vacuum houses, to search for landmines, to service pipelines, to carry out security and surveillance, and many others (see [7] Chapter 9 for a diverse list).

One early approach to robot programming was characterized by the Stanford Research Institute's STRIPS system and the Shakey robot [8]. That approach proceeded as follows: All of Shakey's sensory inputs were first collected. These inputs were then processed in terms of how they related to the physical goals that had been set for Shakey. Once a plan of action had been developed, based on the goals and sensory inputs, signals would be sent to the motors driving Shakey's wheels so as to carry out this plan, step by step. This style of controlling a robot has since come to be called *deliberative* [4]. The central element of the approach involves having the robot deliberate explicitly about the state of its environment and what actions are required to bring this state in line with the robot's goals. It also requires that robot computation proceeds strictly according to a 'Sense, then Plan, then Act' sequence (Figure 9-1(a)). It can be difficult to produce a timely response to sensory input because of this approach. Planning requires modeling the environment as well as the preconditions and effects of robot actions, which can be quite tricky for unstructured environments.

Brooks [6] contrasted this 'vertical' layered approach with a new 'horizontal' layered approach (See Figure 9-1(b)). Sensory information is channeled to all the horizontal layers in parallel, and commands may be sent to the motors from any of the layers. An advantage of this approach is that action can be produced in a timely fashion in response to the sensors, since sensor information does not have to propagate across planning/deliberation layers in order to reach the motors. Another advantage is

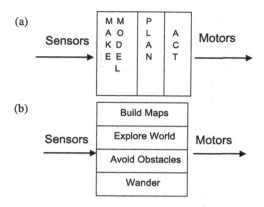

Figure 9-1:   (a) Vertical (deliberative) (b) versus Horizontal (reactive) layered approaches.

that more complex physical behaviors can be created in a modular fashion from simpler behaviors: Additional layers can be added to the architecture by building on top of or 'subsuming' the actions of the simpler behaviors.

Several different robot architectures have been proposed for building behavior-based systems; see Arkin [4] Chapter 4 for a good overview. There are some themes common across many of these architectures:

- Behaviors are seen as concurrent processes. Finite state machines (FSM) have been used by many authors (e.g., [2, 6, 10]) as a specification tool for behaviors.
- Behaviors can be composed to produce more complex behaviors.
- The motor commands of behaviors at different layers may need to be arbitrated before being carried out by the robot's actuators (motors).

Arkin uses Arbib's Schema Theory [3] to build behaviors as networks of perceptual and motor schemas. A *perceptual schema* is a process that continuously extracts the parameters and trigger conditions for a task; a *motor schema* is a process that uses these parameters to carry out the task. A collection of communicating schemas is called a *schema assemblage* and is a coordinated control program for the robot. In some cases, this communication might be the calculation of an input used by another schema, or it might be a synchronization signal to tell another schema to start or to stop. This kind of communication is illustrated in the schema network in Figure 9-2. This figure shows a schema assemblage

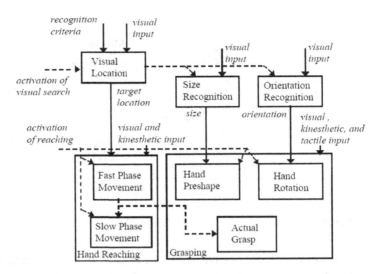

Figure 9-2: Reaching and grasping schema assemblage, after [2].

for reaching, preshaping and grasping an object with a dexterous hand. The communication maps very easily to point-to-point message passing. Lyons and Arbib [12] used port automata [14] as a framework for specifying schema assemblages, an improvement on using FSMs, since concurrency and message passing is explicitly modeled.

There is another kind of communication that needs to happen: when multiple behaviors produce motion commands for the robot, there needs to be *an arbitration between behaviors* or otherwise the actions of the robot may oscillate wildly as it carries out a motion command from one behavior and then potentially a very different command from another behavior. Arbitration allows one behavior to inhibit a second behavior, or to replace the output of the second behavior with a value of its own (Subsumption architecture [6]). It can allow the output commands of two or more behaviors to be linearly summed producing a weighted combination (Motor Schemas [4]) or to be chosen between in priority order (fixed order arbitration [10]).

9.2. Static Behavior-Based Architecture

The behavior-based architecture shown in Figure 9-3 is a fairly standard example based on Arkin's *foraging* problem [4].

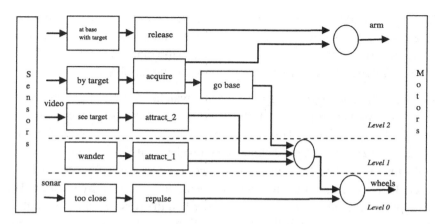

Figure 9-3: Static behavior-based architecture for foraging.

At the lowest layer, layer 0, the robot has an obstacle avoidance behavior, a behavior that moves the robot away from any obstacle that comes too close. The next layer, layer 1, implements wandering and exploring. The top layer, layer 2, implements the foraging behavior, where the robot continually looks for a target object. When found, the target object is acquired and returned to the base location. The boxes in Figure 9-3 are the perceptual and motor schemas for behaviors, the arrows represent the flow of information, and the circles are the arbitrators.

Level 0: The level 0 obstacle avoidance behavior consists of a perceptual schema, *too close*, that uses information from the sonars to determine when an object is too close and sends the direction and distance information to a motor schema. The motor schema builds a *repulsive potential field* around the obstacle. This style of motor schema behavior-based navigation was introduced by Arkin [4]. A virtual repulsive force (with magnitude and direction) is calculated and passed to the level 0 arbitrator, and from there to the motor controller for the wheels, so that robot acts as if it is magnetically repelled by the obstacle and moves away it.

Level 1: The *wander* perceptual schema generates random locations around the robot and passes the direction and distance to a motor schema *attract_1* that generates an attractive virtual potential field around that location. The attractive force (with magnitude and direction) is channeled to the level 1 arbitrator. From there it is directed to the level 0 arbitrator, where it is *vector summed* with the obstacle avoidance information to produce a single force for the motor controller for the wheels. In this way, the robot continues to avoid obstacles while moving to the random locations selected by *wander* and instead directs the robot to the target object.

Level 2: The *see target* perceptual schema watches for the target object and when the target is seen, channels the distance and direction information to an *attract_2* motor schema. This schema calculates a virtual attractive force for the target location, and sends that to the level 1 arbitrator, where it replaces (subsumes) the data from *wander*. Hence, the robot is pulled towards the target location, rather than the random locations generated by *wander*. The perceptual schema *by target* is triggered when the robot is close enough to acquire the target. It sends the object location to the *acquire* motor schema. The *acquire* schema directs the grasping and acquisition of the target object. This itself might be a schema assemblage (e.g., Figure 9-2).

When grasping is complete, a signal is sent to the *go base* motor schema which calculates a virtual attractive potential field around the base location. The attractive field information is channeled to the level 1 arbitrator where it subsumes both the *wander* and the target *attract* data, forcing the robot to be pulled towards its base.

The *at base with target* perceptual schema is triggered when the robot returns to the base with the target object. It also triggers the *release* motor schema that directs the arm to release the object at the base. Notice that the *acquire* and *release* motor schemas both control the arm and are channeled therefore through an arbitrator. The schema *release* will subsume *acquire* so that robot does not try to reacquire target objects left at the base.

9.2.1. *Partitioning*

We will view each schema as a port automaton [14] — a state machine that can communicate with the other schemas over a set of communication ports. We can also view the arbitrators as port automata. Each automaton can be mapped to a processor in our cluster. There are two partition questions we then need to address: how to provide the code for each automaton to the processor and how to implement communication on ports between the automata.

(1) Our use of MPI until now has been restricted to the SPMD (Single Program/process Multiple Data) model, and all processors had basically

the same instructions. The programs we developed tested the rank variable to see if they were running on a root or some other processor and only then executed specialized instructions. Using this scheme to implement our port automata would be difficult: we would have a program that is essentially a large case statement that tested the rank variable and executed different code for each automaton.

However, MPI also allows a Multiple Program/process Multiple Data model (MPMD) in which different programs can be specified for different processors, but in which all processors remain in the same communicator, MPI_COMM_WORLD. This allows us to write separate code for each automaton and still have them communicate in the standard way.

(2) Point to point communicate operations (MPI_Send and MPI_Recv) will allow our automata on each processor to communicate. However, MPI communication has no concept of a port — a communication channel over which multiple messages can be sent. We can use the tag field of the communication to implement named ports. Specifying the tag field in an MPI_Recv command will allow messages sent on the named port to be preferentially read before any messages on another port.

9.2.2. *Program Design*

Each schema will be a separate program, with separate MPI_Init and MPI_Finalize operations. In the case that we have programs *schema_1*, *schema_2*, . . . , *schema_n* then these can be executed on the cluster with the command:

```
mpirun −np 1 schema_1 : -np 1 schema_2 : ... : -np 1 schema_n
```

These programs will be assigned ranks consecutively, and they will all be in the communicator MPI_COMM_WORLD. Collective and point-to-point communications can be carried out between them. In this static behavior-based architecture, the assignment of ranks to schemas is static and can be encoded in a global rank (schema) look-up function for use in message passing function arguments.

In the schema MPI details below, will assume each schema is a separate program with its own MPI_Init and MPI_Finalize and just discuss the internal loop for the schema, where it reads sensory or port information and produces output port information.

Schema: too-close. The range information from each of the sonar sensors is read, and if an obstacle is considered too close to the robot, that sonar range and (fixed) bearing information is written to the schema's output ports. If nothing is too close, a special message is sent anyway.

```
sentMessage=false;
for ( s = 0; s < NUM_SONARS; s++ )
    if (sonar_range[s] < MIN_RANGE) {
        MPI_Send(&sonar_range[s], 1, MPI_INT, rank(repulse), PORT-R,
                MPI_COMM_WORLD);
        MPI_Send(&sonar_angle[s], 1, MPI_FLOAT, rank(repulse),
                PORT-B, MPI_COMM_WORLD);
        sentMessage=true;
    }
if ( !sentMessage ) { /* special 'no obstacle' message */
        MPI_Send(&maxRange, 1, MPI_INT, rank(repulse), PORT-R,
                MPI_COMM_WORLD);
        MPI_Send(&noBearing, 1, MPI_FLOAT, rank(repulse), PORT-B,
                MPI_COMM_WORLD);
    }
```

Schema: repulse. When range and bearing information is received, the effect of a repulsive potential field placed at that range and bearing is calculated and transmitted. The repulsive field is proportional to the inverse square of the distance to the obstacle.

```
MPI_Recv(&range, 1, MPI_INT, rank(too-close), PORT-R,
        MPI_COMM_WORLD);
MPI_Recv(&bearing, 1, MPI_FLOAT, rank(too-close), PORT-B,
        MPI_COMM_WORLD);
forceMag = gainR / (range*range) ; /* inverse square */
forceDir = pi + bearing; /* pointing away from obstacle */
MPI_Send(&forceMag, 1, MPI_FLOAT, rank(level-0-arbit), PORT-MR0,
        MPI_COMM_WORLD);
MPI_Send(&forceDir, 1, MPI_FLOAT, rank(level-0-arbit), PORT-DR0,
        MPI_COMM_WORLD);
```

Schema: wander. A random distance, bearing and time is generated and then continually transmitted for a random duration. We will assume that random generates a random number between 0 and its argument.

```
range = random(MAX_D); bearing = random(2*pi);
duration = random(MAX_T); now = time();
while ( time() < now+duration ) {
    MPI_Send(&range, 1, MPI_FLOAT, rank(attract_1), PORT-M0,
            MPI_COMM_WORLD);
    MPI_Send(&bearing, 1, MPI_FLOAT, rank(attract_1), PORT-D0,
            MPI_COMM_WORLD);
    Sleep(smallTime);
}
```

Schema: attract_1. When the range and bearing information is received, the effect of an attractive potential field placed at the range and bearing is calculated and transmitted. The attractive field is calculated as a constant force everywhere except close to the target location, where it decreases proportional to distance. Since *wander* transmits a fixed range, the result is a fixed attractive force.

```
MPI_Recv(&range, 1, MPI_INT, rank(wander), PORT-R,
        MPI_COMM_WORLD);
MPI_Recv(&bearing, 1, MPI_FLOAT, rank(wander), PORT-B,
        MPI_COMM_WORLD);
forceMag = MIN(MAX_A, gainA*range); /* ballistic/proportional rule*/
forceDir = bearing; /* pointing towards the target */
MPI_Send(&forceMag, 1, MPI_FLOAT, rank(level-1-arbit), PORT-MA0,
        MPI_COMM_WORLD);
MPI_Send(&forceDir, 1, MPI_FLOAT, rank(level-1-arbit), PORT-DA0,
        MPI_COMM_WORLD);
```

Schema: level-0-arbit. The vector sum of the attractive and repulsive forces is calculated and transmitted to the motors.

```
MPI_Recv(&fMA, 1, MPI_INT, rank(level-1-arbit), PORT-M,
        MPI_COMM_WORLD);
MPI_Recv(&fMR, 1, MPI_INT, rank(repulse), PORT-MR0,
        MPI_COMM_WORLD);
MPI_Recv(&fDA, 1, MPI_FLOAT, rank(level-1-arbit), PORT-D,
        MPI_COMM_WORLD);
MPI_Recv(&fDR, 1, MPI_FLOAT, rank(repulse), PORT-DR0,
        MPI_COMM_WORLD);
fM = ( fMA + fMR ) / 2.0;
fD = ( fDA  + fBR ) / 2.0; /* sum magnitude and direction */
setRobotAngle(fD);
setRobotVelocity(gainV*fM);
```

Schema: see-target. The target is identified as a location in the camera image, and the location is transformed from image to world coordinates using the camera calibration information. If the target is not recognized in the camera image, no data is sent. The output is directed to the second instance of the *attract* schema.

```
getCameraImage( Image );
result = recognizeTarget( Image, targetModel );
if ( result.found ) {
mapImagetoWorldCoordinates( result.location,
                            target.location, cameraCalib );
  bearing = atan2(target.location.y, target.location.x);
  range = sqrt( sqr(target.location.x) + sqr(target.location.y) );
  MPI_Send(&range, 1, MPI_FLOAT, rank(attract_2), PORT-R,
          MPI_COMM_WORLD);
  MPI_Send(&bearing, 1, MPI_FLOAT, rank(attract_2), PORT-B,
          MPI_COMM_WORLD);
}
```

Schema: attract_2. This is the same as *attract_1* except that it receives its input from *see-target* and sends its output to the level 1 arbitrator on the ports **PORT-DA1** and **PORT-MA1**.

Schema: level-1-arbit. At this point, the level 1 arbitrator has two attract schemas connected to it: one on **PORT-MA0** and **PORT-DA0** and one on **PORT-MA1** and **PRT-DA1**. There will be another connection from the *go* *base* schema. The higher port numbers *have priority over* the lower if they carry data. The MPI_Iprobe command allows us to implement this priority arbitration:

```
int MPI_Iprobe(    int          source,
                   int          tag,
                   MPI_Comm     comm,
                   int          *flag,
                   MPI_Status   *status ).
```

This command tests to see if there is a message from source waiting on the port tag and returns the result in flag. The arbitrator loops, testing each port in order of priority, receiving only the highest priority message and passing it on to its output ports. To allow this to be written as a loop, the ranks and port names are initialized into arrays schemaRank and portNames in inverse priority order. So schemaRank[0] contains rank(attract_1), portNames[0][0] contains PORT-MA0, and port-Names[0][1] contains PORT-DA0 .

```
for ( p = NUM_CONNECTIONS-1; p >= 0; p-- ) { /* scan high to low */
  MPI_Iprobe( schemaRank[p],  portName[p][0],
              MPI_COMM_WORLD, &messageWaiting, &status);
  if ( messageWaiting ) { /* assume both ports can now communicate*/
    MPI_Recv(&mag, 1, MPI_FLOAT, schemaRank[p],
             portName[p][0], MPI_COMM_WORLD);
    MPI_Recv(&dir, 1, MPI_FLOAT, schemaRank[p], portName[p][1],
             MPI_COMM_WORLD);
    MPI_Send(&mag, 1, MPI_FLOAT, rank(level-0-arbit), PORT-M,
             MPI_COMM_WORLD);
    MPI_Send(&dir, 1, MPI_FLOAT, rank(level-0-arbit), PORT-D,
             MPI_COMM_WORLD);
  }
}
```

9.2.3. *Analysis*

All the communication operations in this architecture are *synchronous* (except for the level 1 arbitrator, to which we'll come back). In MPI what this means is that after a send operation has occurred on a port, a second send can't happen until the first is read. Once a network of schemas has started synchronous communications, then the input and output operations of connected schemas are overlapped in time in a lockstep fashion. The parallel execution time for the network is therefore the same as the execution time for the schema that takes the longest to execute:

$$T_p = \max_i(T_{\text{schema }i})$$

The implication of this is that a slower level 2 behavior (such as the schema for recognizing a visual target) will slow down the much faster layer 0 behaviors (such as obstacle avoidance). This slowdown will happen when the arbitrator needs to wait for inputs from the different layers before making its arbitration decision. To avoid this problem, the level 1 arbitrator schema uses an *asynchronous* operation to test whether there is input available before proceeding.

A different solution was used for the level 0 arbitrator: in that case, the *too close* schema sends a message whether it sees an obstacle or not. The "no obstacle" message produces a negligible repulsive effect. An asynchronous operation would have allowed us to do this without the "no obstacle" message.

However, by introducing asynchrony we also introduce timing problems. Consider the case where a target is seen, and the *attract_2* motor schema produces a signal that subsumes the *attract_1* wandering behavior signal. Since the loop time of *see target* is large compared to *wander*, a subsequent new message will arrive quickly from *attract_1*. Even though this is a lower priority signal — it's the only one the arbitrator sees, so it will get selected. The emergent behavior is of the robot alternately wandering and moving to the target (if it's still visible).

Adding a delay in the loop of the level 1 arbitrator schema will reduce the problem but slow everything down to layer 2 execution times again. A better solution is to modify *attract_2* so that it generates multiple output messages for each input message. The *attract_2* schema should use MPI_Iprobe to test its input, and as long as there is no input, produce the same output as the last input. In that case, however, *see target* needs to tell *attract_2* whether the target has been seen or not, so it can stop transmitting if no target is visible. This modification introduces a hysteresis so that wandering is disabled for as long as the target is seen.

Schema: see-target.

```
getCameraImage( Image );
result = recognizeTarget( Image, targetModel );
MPI_Send(&result.found, 1, MPI_INT, rank(attract_2), PORT-T,
        MPI_COMM_WORLD);
if (result.found) {
mapImagetoWorldCoordinates( result.location,
                            target.location, cameraCalib );
bearing = atan2( target.location.y, target.location.x );
range = sqrt( sqr(target.location.x) + sqr(target.location.y) );
MPI_Send(&range, 1, MPI_FLOAT, rank(attract_2), PORT-R,
        MPI_COMM_WORLD);
MPI_Send(&bearing, 1, MPI_FLOAT, rank(attract_2), PORT-B,
        MPI_COMM_WORLD);
}
```

Schema: attract_2.

```
MPI_Iprobe( rank(see-target), PORT-T, MPI_COMM_WORLD,
           &flag, &status);
if (flag) { /* update information with latest*/
    MPI_Recv(&found, 1, MPI_INT, rank(see-target), PORT-T,
            MPI_COMM_WORLD);
    MPI_Recv(&range, 1, MPI_INT, rank(see-target), PORT-R,
            MPI_COMM_WORLD);
    MPI_Recv(&bearing, 1, MPI_FLOAT, rank(see-target), PORT-B,
            MPI_COMM_WORLD);
}
If (found) { /* act on last data received */
    forceMag = MIN(MAX_A, gainA*range); /* ballistic/proportional rule*/
    forceDir = bearing; /* pointing towards the target */
    MPI_Send(&forceMag, 1, MPI_FLOAT, rank(level-1-arbit), PORT-
            MA1, MPI_COMM_WORLD);
    MPI_Send(&forceDir, 1, MPI_FLOAT, rank(level-1-arbit), PORT-DA1,
            MPI_COMM_WORLD);
}
```

9.3.   Dynamic Behavior-Based Architecture

The obstacle avoidance behavior in layer 0 has the flaw that if there are two or more different obstacles seen by *too-close* then their effect will be sent consecutively to the level 0 arbitrator. The emergent behavior is that the robot will be pushed away by one obstacle first and then by the second. It's possible that these inputs can be integrated by the motor controller to produce less oscillatory behavior but it would be better if the perceptual schema *too-close* handled multiple obstacles better. Since there are 16 sonars, we could make 16 *too-close* schemas, each of which tested one sensor. This wouldn't scale as well for laser ranging, however.

Another approach is to generalize our port-automaton framework to allow for the dynamic creation and destruction of automata. This approach to the specific of behavior-based system is described in [11, 13]. The dynamic approach has several advantages:

(1) It doesn't create schemas and use computational resources unless they are needed. For example, in Figure 9-3 the *repulse schema* always calculates a repulsive field, but it's only useful when there actually is an object close. Similarly, *attract_1*, *attract_2*, etc., always calculate output messages whether or not their inputs are triggered or their outputs necessary.

(2) It can create as many schemas as needed. For example, the *too-close* schema could create *repulse* schemas for each obstacle detected.

Behavior-based architectures do not typically have dynamic processes. They are typically composed of collections of static modules, as was the foraging architecture of the previous section. Therefore, we need to introduce some additional notation in order to represent the dynamic aspect of the architecture.

A process algebra notation is used in [11, 13] for dynamic process networks. That allows the clear indication of process creation, communications and termination. However, here we just adopt a minor change in notation to Figure 9-3, which uses a fairly standard behavior-based notation. We will use a *zig-zag arrow* to indicate the dynamic creation

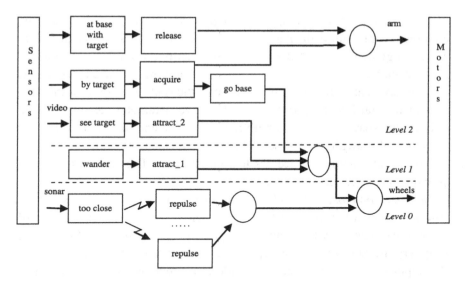

Figure 9-4:   Dynamic behavior-based architecture for foraging.

of a schema. Figure 9-4 shows level 0 rewritten with this convention: the schema *too close* now dynamically creates new perceptual schemas to monitor each obstacle detected and vector sums their effect as the repulsive input for the level 0 arbitrator.

9.3.1.   *Program Design*

New processes can be created and connected dynamically in MPI using the MPI_Comm_spawn command:

```
int MPI_Comm_spawn(    char        *command,
                       char        *argv[],
                       int         maxprocs,
                       MPI_Info    info,
                       int         root,
                       MPI_Comm    comm,
                       MPI_Comm    *intercomm,
                       int         errcodes[ ]    ).
```

Each new process is mapped to a processor and given a rank. The command parameter is the name of the executable to be spawned and argv are the arguments to it. maxprocs is the maximum number of copies of this executable to make (similar to the "-np" argument of mpirun). Info is a set of key-value pairs with information to the runtime system of the MPI system for starting the process. The first five parameters are interpreted only for the spawning process with rank root. The communicator comm is the group of spawning processes. The communicator intercom allows both spawned and spawning processes to communicate. The array errcodes is contains one code for each spawned process.

Schema: too close. This schema will check the sonar sensors as before. But whenever a sensor reports a range that is too small, an instance of the *repulse* schema will be made to represent that sensor. The first step is to determine how many *new* repulse schema instances need to be made; if a repulse schema instance was made in a previous cycle but the range sensor still returns a small range, the repulse schema instance will still be operating as a child process.

```
numToCreate=0;
for ( s = 0; s < NUM_SONARS; s++ )
    If ( sonar_range[ s ] < MIN_RANGE && !repulseCreated[ s ] ) {
        /* a repulse is needed */
        repulseCreated[ s ] = true; /* so that another is not created*/
        sprintf(args[ numToCreate ], "%d", s);
        /* add sensor index to arg list for spawn*/
        numToCreate++; numCreated++;
    }
    else {
        repulseCreated[ s ] = false; /* this repulse will have terminated */
        numCreated--;
    }
```

The args list is the list of range sensors that produced a short range, and are not yet being monitored by *repulse* schema instances. numToCreate is the number of such sensors, and numCreated is the total number of *repulse* schema instances created so far.

```
/* spawn the new repulse schema instances */
MPI_Bcast(&numToCreate,1,MPI_INT,mergedcomm);
/* alert child processes */
If ( numToCreate > 0 ) { /* need to spawn new repulse schema instances */
        MPI_Comm_spawn("repulse", args, numToCreate, MPI_INFO_NULL, 0,
                        MPI_COMM_SELF, &intercom,
                        MPI_ERRCODES_IGNORE );
        MPI_Intercomm_merge(intercom, 0, &mergedcomm);
        /* include new ranks */
}
```

To allow any previously created repulse schema instances to communicate with the new ones created in this cycle (if any), those processes need to also execute a collective MPI_Comm_spawn. The broadcast command alerts them to this information so they participate in the collective command if it is executed. Once the child processes are created, the parent and child communicators are merged so that all can communicate. The new communicator is called mergedcomm. The next step is to collect the information from all *repulse* schema instances using a collective reduction operation.

```
/* get the repulsive force contribution from each repulse summed */
local_f[0]=local_f[1]=0.0;
MPI_Reduce(&local_f, &f, 2, MPI_FLOAT, MPI_SUM, root,
                mergedcomm);
f[0] /= numCreated; f[1] /= numCreated; /* average the repulsive force */
MPI_Send(&f[0], 1, MPI_FLOAT, rank(level-0-arbit), PORT-MR0,
                MPI_COMM_WORLD);
MPI_Send(&f[1], 1, MPI_FLOAT, rank(level-0-arbit), PORT-DR0,
                MPI_COMM_WORLD);
```

The reduction operation sums the magnitude and direction information from each repulse schema instance on the root. This information is averaged and then communicated to the level 0 arbitrator.

Schema: repulse. This schema is now responsible for checking the sonar input and for passing on information as long as there is an obstacle present. It terminates itself when there is no obstacle present. The first step is to establish the correct communicator. In the case that the process was just created, then the parent and children communicators need to be merged. In the case that the process exists already, it needs to participate *as a parent* in the creation of any new processes to update it's communicator.

```
If ( justCreated ) { /* new instance of repulse */
    MPI_Comm_get_parent(&newcomm); /* get parents communicator */
    /* merge  parent and child ranks */
    MPI_Intercomm_merge(newcomm, 0,  &mergedcomm);
}
else { /* not a new instance of repulse */
    MPI_Bcast(&numToCreate, MPI_INT,mergedcomm);
    /* alert child processes */
    if (numToCreate>0) { /* participate in collective spawn */
        MPI_Comm_spawn("",0,0,MPI_INFO_NULL,0,mergedcomm,&newcomm,
                            MPI_ERRCODES_IGNORE);
        /* merge parent and child ranks */
        MPI_Intercomm_merge(newcomm, 0,  &mergedcomm);
    }
}
```

Having established the correct communicator, *repulse* can calculate the repulsive field component caused by the sensor it received as argument **mySensor**. If the range sensor shows nothing, this worker process terminates, otherwise it participates in the collective reduction.

```
range = sonar_range[ mySensor ];
bearing = sonar_angle[ mySensor ];
If ( range  >=  MIN_RANGE ) { /* terminate the process */ }
else {
    local_f[0] = gainR / (range*range) ; /* inverse square */
    local_f[1] = pi + bearing; /* pointing away from obstacle */
    MPI_Reduce( &local_f, &f, 2, MPI_FLOAT, MPI_SUM,
                        root, mergedcomm);
}
```

9.3.2. *Analysis*

The dynamic behavior-based architecture implementation comes at the cost of three additional collective operations: spawn, broadcast and reduce.

Even though the cost of the reduction operation grows more slowly than the cost of the multiple point to point operations that would be necessary (in the arbitrator) if we just extended the static architecture with multiple repulse schema instances, the dynamic architecture will typically have higher communication costs than the equivalent static architecture.

On the other hand, the dynamic architecture has better processor utilization: the static architecture needs to provide repulse schemas for the worst case. However, in all other situations, it will underutilize the processors dedicated to these schemas. Since the dynamic architecture only creates repulse schemas as needed there is no underutilization cost.

9.4. Summary

In this chapter we have switched from looking at robot algorithms to looking at robot architectures. A robot architecture is typically a software architecture. In particular we looked at behavior-based architectures, where the robot program is built as a collection of mostly independent, communicating modules responsible for physical robot behaviors. We relied heavily on Arkin's [4] motor schema approach to behavior based architectures, using the Lyons and Arbib [12] port automaton model for perceptual and motor schemas.

Most behavior-based architectures are static, hierarchical networks, with behaviors in higher levels modifying the behavior in lower levels. This modification happens with the use of arbitrators such as Arkin's vector summation [4], Brook's inhibition and suppression [6], or Jones fixed order arbitration [10].

The problems we have looked at to this point consist of a single body of code with some conditionals to control the behavior of different ranks. The Single Program/process Multiple Data (SPMD) module of MPI worked well for this. Building a behavior-based architecture demanded that we have a number of very different code modules: one for each behavior. We needed to switch to using the Multiple Program/process Multiple Data (MPMD) functionality of MPI to handle this.

We used the MPMD functionality to build a static architecture for the foraging problem. Schemas used the MPI point-to-point communication commands to communicate. The tag fields were used to identify the communication ports over which schemas communicated. However, a network of synchronous point-to-point communications raised issues since it forced the faster, lower layer behaviors to slow down to the same rate as the slower, upper layer behaviors. We addressed this by introducing asynchronous communication at selected points.

The obstacle avoidance behavior introduced for the foraging example had a limited capacity to handle multiple obstacles. More static capacity could be added, by adding more schemas. A more efficient use of processing power would be to dynamically create schemas to handle obstacles as needed.

We constructed a dynamic behavior-based architecture for the lowest level of the foraging problem. The process spawning functionality of MPI was used to create new schemas to handle the workload of monitoring obstacles and calculating their repulsive potential field. In fact, this is a fairly standard parallel programming paradigm with just one novel twist: a set of worker processes are created that divide the problem between them, with the additional twist that the set is dynamic. As obstacles go from view the worker processes representing them terminate, and as new obstacles are seen, new worker processes are added.

References

1. Agre, P. and Chapman, D., PENGI: An implementation of a theory of activity. In *Proceedings of the Sixth National Conference on Artificial Intelligence (AAAI-87)*, pp. 268–272, Seattle WA (1987).
2. Arbib, M.A., Iberall, T., and Lyons, D.M., Co-ordinated Control Programs for Movements of the Hand, *Exp. Brain Res. #10*, Springer-Verlag 1985.
3. Arbib, M.A., *The Metaphorical Brain 2*, Wiley & Sons 1989.
4. Arkin, R.C., *Behavior-Based Robotics*, MIT Press 1998.
5. Bekey, G.A., *Autonomous Robots*, MIT Press 2005.
6. Brooks, R.A., A robust layered control system for a mobile robot. *IEEE Journal of Robotics and Automation*, 2(1):14–23, 1986.
7. Dudek, G. and Jenkin, M., *Computational Principles of Mobile Robotics*, Cambridge Press 2000.
8. Fikes, R.E. and Nilsson, N., STRIPS: A new approach to the application of theorem proving to problem solving. *Artificial Intelligence*, 5(2):189–208, 1971.
9. Firby, J.A., An investigation into reactive planning in complex domains. In *Proceedings of the Tenth International Joint Conference on Artificial Intelligence (IJCAI-87)*, pp. 202–206, Milan Italy, 1987.
10. Jones, J., *Robot Programming: A Practical guide to Behavior-Based Robotics*, McGraw-Hill 2004.

11. Lyons, D.M., Representing and Analyzing Action Plans as Networks of Concurrent Processes, *IEEE Transactions on Robotics and Automation*, 9(3), June 1993.

12. Lyons, D.M. and Arbib, M.A., A Formal Model of Computation for Sensory-based Robotics, *IEEE Transactions on Robotics and Automation* 5(3), June 1989.

13. Lyons, D.M. and Arkin, R., Towards Performance Guarantees for Emergent Behavior, *IEEE International Conference on Robotics and Automation*, New Orleans LA, 2004.

14. Steenstrup, M., Arbib, M.A., and Manes, E.G., Port automata and the algebra of concurrent process, *J. Computer System Sciences* 27(1):29–50, 1983.

Appendix I: Summary of OpenMPI Man Page for mpirun

Single Process Multiple Data (SPMD) Model:

mpirun [options] **\<program>** [\<args>]

Multiple Instruction Multiple Data (MIMD) Model:

mpirun [global_options]   [local_options1] **\<program1>** [\<args1>] :
[local_options2] **\<program2>** [\<args2>] : ... :
[local_optionsN] **\<programN>** [\<argsN>]

Description:

One invocation of *mpirun* starts an MPI application running under Open MPI. If the application is single process multiple data (SPMD), the application can be specified on the *mpirun* command line. If the application is multiple instruction multiple data (MIMD), comprising of multiple programs, the set of programs and argument can be specified in one of two ways: Extended Command Line Arguments and Application Context. Extended command line arguments allow for the description of the application layout on the command line using colons (:) to separate the specification of programs and arguments. Some options are globally set across all specified programs (e.g., -hostfile), while others are specific to a single program (e.g., -np).

Specifying Host Nodes:

Host nodes can be identified on the *mpirun* command line with the *-host* option or in a hostfile.

For example,

```
mpirun -H aa, aa, bb ./a.out
```

launches two processes on node aa and one on bb.
Or, consider the hostfile

```
% cat myhostfile
aa slots=2
bb slots=2
cc slots=2
```

Here, we list both the host names (aa, bb, and cc) but also how many "slots" there are for each. Slots indicate how many processes can potentially execute on a node. For best performance, the number of slots may be chosen to be the number of cores on the node or the number of processor sockets. If the hostfile does not provide slots information, a default of 1 is assumed. When running under resource managers (e.g., SLURM, Torque, etc.), Open MPI will obtain both the hostnames and the number of slots directly from the resource manger.

```
mpirun -hostfile myhostfile ./a.out
```

will launch two processes on each of the three nodes.

```
mpirun -hostfile myhostfile -host aa ./a.out
```

will launch two processes, both on node aa.

Specifying Number of Processes:

As we have just seen, the number of processes to run can be set using the hostfile. Other mechanisms exist. The number of processes launched can be specified as a multiple of the number of nodes or processor sockets available. For example,

```
mpirun -H aa, bb -npersocket 2 ./a.out
```

launches processes 0–3 on node aa and process 4–7 on node bb, where aa and bb are both dual-socket nodes. The *-npersocket* option also turns on the *-bind-to-socket* option.

Another alternative is to specify the number of processes with the *-np* option. Consider now the hostfile

```
% cat myhostfile
aa slots=4
bb slots=4
cc slots=4

mpirun -hostfile myhostfile -np 6 ./a.out
```

will launch ranks 0–3 on node aa and ranks 4–5 on node bb. The remaining slots in the hostfile will not be used since the *-np* option indicated that only 6 processes should be launched.

Mapping Processes to Nodes: Using Policies:

The examples above illustrate the default mapping of process ranks to nodes. This mapping can also be controlled with various *mpirun* options that describe mapping policies. Consider the same hostfile as above, again with *-np* 6:

	node aa	node bb	node cc
mpirun	0 1 2 3	4 5	
mpirun -loadbalance	0 1	2 3	4 5
mpirun -bynode	0 3	1 4	2 5
mpirun -nolocal		0 1 2 3	4 5

The *-loadbalance* option tries to spread processes out fairly among the nodes. The *-bynode* option does likewise but numbers the processes in "by node" in a round-robin fashion.

The *-nolocal* option prevents any processes from being mapped onto the local host (in this case node aa). While *mpirun* typically consumes few system resources, *-nolocal* can be helpful for launching very large jobs where *mpirun* may actually need to use noticeable amounts of memory and/or processing time. Just as *-np* can specify fewer processes than there are slots, it can also oversubscribe the slots. For example, with the same hostfile:

mpirun -hostfile myhostfile -np 14 ./a.out

will launch processes 0–3 on node aa, 4–7 on bb, and 8–11 on cc. It will then add the remaining two processes to whichever nodes it chooses.

Limits to oversubscription can be specified in the hostfile itself:

```
% cat myhostfile
aa slots=4 max_slots=4
bb max_slots=4
cc slots=4
```

Of course, *-np* can also be used with the *−H* or *-host* option. For example,

```
mpirun -H aa,bb -np 8 ./a.out
```

launches 8 processes. Since only two hosts are specified, after the first two processes are mapped, one to aa and one to bb, the remaining processes oversubscribe the specified hosts.

And here is a MIMD example:

```
mpirun -H aa -np 1 hostname : -H bb,cc -np 2 uptime
```

will launch process 0 running *hostname* on node aa and processes 1 and 2 each running *uptime* on nodes bb and cc, respectively.

Process Binding:

Processes may be bound to specific resources on a node. This can improve performance if the operating system is placing processes suboptimally. For example, it might oversubscribe some multi-core processor sockets, leaving other sockets idle; this can lead processes to contend unnecessarily for common resources. Or, it might spread processes out too widely; this can be suboptimal if application performance is sensitive to interprocess communication costs. Binding can also keep the operating system from migrating processes excessively, regardless of how optimally those processes were placed to begin with.

To bind processes, one must first associate them with the resources on which they should run. For example, the *-bycore* option associates the processes on a node with successive cores. Or, *-bysocket* associates the processes with successive processor sockets, cycling through the sockets in a round-robin fashion if necessary. And *-cpus-per-proc* indicates how many cores to bind per process.

But, such association is meaningless unless the processes are actually bound to those resources. The binding option specifies the granularity of binding — say, with *-bind-to-core* or *-bind-to-socket*. One can also turn binding off with *-bind-to-none*, which is typically the default.

Finally, *-report-bindings* can be used to report bindings.

As an example, consider a node with two processor sockets, each comprising four cores. We run *mpirun* with *-np 4 -report-bindings* and the following additional options:

```
% mpirun ... -bycore -bind-to-core
[...] ... binding child [...,0] to cpus 0001
[...] ... binding child [...,1] to cpus 0002
[...] ... binding child [...,2] to cpus 0004
[...] ... binding child [...,3] to cpus 0008
% mpirun ... -bysocket -bind-to-socket
[...] ... binding child [...,0] to socket 0 cpus 000f
[...] ... binding child [...,1] to socket 1 cpus 00f0
[...] ... binding child [...,2] to socket 0 cpus 000f
[...] ... binding child [...,3] to socket 1 cpus 00f0
% mpirun ... -cpus-per-proc 2 -bind-to-core
[...] ... binding child [...,0] to cpus 0003
[...] ... binding child [...,1] to cpus 000c
[...] ... binding child [...,2] to cpus 0030
[...] ... binding child [...,3] to cpus 00c0
% mpirun ... -bind-to-none
```

Here, *-report-bindings* shows the binding of each process as a mask. In the first case, the processes bind to successive cores as indicated by the masks 0001, 0002, 0004, and 0008. In the second case, processes bind to all cores on successive sockets as indicated by the masks 000f and 00f0. The processes cycle through the processor sockets in a round-robin fashion as many times as are needed. In the third case, the masks show us that 2 cores have been bind per process. In the fourth case, binding is turned off and no bindings are reported.

Open MPI's support for process binding depends on the underlying operating system. Therefore, processing binding may not be available on every system.

Appendix II: MPI Datatypes

Table AII-1: Table of MPI and equivalent C datatypes from MPI Interface Standard Version 2.2, 2009.

MPI Datatype	C Datatype
MPI_CHAR	char (treated as printable character)
MPI_SHORT	signed short int
MPI_INT	signed int
MPI_LONG	signed long int
MPI_LONG_LONG_INT	signed long long int
MPI_LONG_LONG	(as a synonym) signed long long int
MPI_SIGNED_CHAR	signed char (treated as integral value)
MPI_UNSIGNED_CHAR	unsigned char (treated as integral value)
MPI_UNSIGNED_SHORT	unsigned short int
MPI_UNSIGNED	unsigned int
MPI_UNSIGNED_LONG	unsigned long int
MPI_UNSIGNED_LONG_LONG	unsigned long long int
MPI_FLOAT	float
MPI_DOUBLE	double
MPI_LONG_DOUBLE	long double
MPI_WCHAR	wchar_t (defined in <stddef.h>)
MPI_C_BOOL	_Bool
MPI_INT8_T	int8_t
MPI_INT16_T	int16_t
MPI_INT32_T	int32_t
MPI_INT64_T	int64_t
MPI_UINT8_T	uint8_t
MPI_UINT16_T	uint16_t
MPI_UINT32_T	uint32_t
MPI_UINT64_T	uint64_t

(Continued)

Table AII-1: *(Continued)*

MPI Datatype	C Datatype
MPI_C_COMPLEX	float _Complex
MPI_C_FLOAT_COMPLEX	(as a synonym) float _Complex
MPI_C_DOUBLE_COMPLEX	double _Complex
MPI_C_LONG_DOUBLE_COMPLEX	long double _Complex
MPI_BYTE	
MPI_PACKED	

Appendix III: MPI Reduction Operations

Table AIII-1: Table of MPI reduction operations from MPI Interface Standard Version 2.2, 2009.

Name	Meaning
MPI_MAX	maximum
MPI_MIN	minimum
MPI_SUM	sum
MPI_PROD	product
MPI_LAND	logical and
MPI_BAND	bit-wise and
MPI_LOR	logical or
MPI_BOR	bit-wise or
MPI_LXOR	logical exclusive or (xor)
MPI_BXOR	bit-wise exclusive or (xor)
MPI_MAXLOC	max value and location
MPI_MINLOC	min value and location

New reduction operations can be defined using the MPI function:

```
int MPI_Op_create(   MPI        User_function *function,
                     int        commute,
                     MPI_Op     *op                        ).
```

The operation must be associative, and if the commute argument is set, then it is assumed to be commutative as well. The user defined function must have the following signature:

```
void function(     void         *invec,
                   void         *inoutvec,
                   int          *len,
                   MPI_Datatype *datatype          ).
```

The arguments invec and inoutvec are arrays of length len that are combined by the function. The result is written to inoutvec.

Appendix IV: MPI Application Programmer Interface

List of MPI functions by order of Chapter references.

The number after each MPI API function is the item number in the alphabetical list.

Chapter 3:

```
MPI_Init();              11
MPI_Comm_rank();          4
MPI_Comm_size();          5
MPI_Finalize();           9
```

Chapter 4:

```
#define BLOCK_LOW(id,p,n)  ((id) * (n) / (p))
#define BLOCK_HIGH(id,p,n) (BLOCK_LOW((id)+1,p,n) – 1 )
#define BLOCK_SIZE(id,p,n)  (BLOCK_HIGH(id,p,n) – BLOCK_LOW(id,p,n) + 1)
#define BLOCK_OWNER(index,p,n) ( ( (p)*( (index) + 1 )  – 1 ) / (n))
```

```
MPI_Scatter ();          15
MPI_Reduce ( );          14
MPI_Gather ();           10
```

Chapter 5:

```
MPI_Bcast ();             2
MPI_Scatterv();          16
```

Chapter 6:

Chapter 8:

Chapter 9:

Alphabetic list of MPI functions used in the text and their descriptions.

1. int MPI_Allreduce (void   *sendbuf,
                        void   *recvbuf,
                        int   count,
                        MPI_Datatype datatype,
                        MPI_Op   op,
                        MPI_Comm   comm   )

MPI_Allreduce performs a *reduction operation* on data from every processor and send the result of the operation back *to every processor*. It is a collective communication function — that is, all the processors in the communicator (MPI_COMM_WORLD in general) participate in the operation. A reduction operation is a binary associative operation (these include sum and product, but also maximum, minimum and logical operations; see Appendix III for a full list of MPI's built-in reduction operations). The parameter sendbuf is the address of the local subtotal and recvbuf is the address of the final reduced total. If a list (array) of items is being separately reduced, then count can be used to indicate the size of the array. The reduction operation is specified by the value of op.

2. int MPI_Bcast (void   *buffer,

                       int   count,

                       MPI_Datatype datatype,

                       int   root,

                       MPI_Comm   comm   )

The broadcast operation **MPI_Bcast** sends data to all processors. It is a collective communication function — that is, all the processors in the communicator (**MPI_COMM_WORLD** in general) participate in the operation. The **buffer** parameter is the address of the array (buffer) being broadcast. The **count** parameter contains the size of the block to be sent and **datatype** is the type of data elements in the block. All count values in the **buffer** on the root processor are transmitted to the local copy of **buffer** on each processor.

3. int MPI_Cart_create (MPI_Comm   comm_old,

                            int   ndims,

                            int   *dims,

                            int   *periods,

                            int   reorder,

                            MPI_Comm   *comm_cart   )

MPI_Cart_create creates a new communicator, **comm_cart**, to send messages within the Cartesian grid processor topology we have created. The **ndims** parameter is the number of dimensions and the **dims** array is the array of processor sizes for each dimension. The array period (set to 1 or 0 for each dimension) controls whether that dimension wraps around. The reorder flag controls whether the processor ranks in the new communicator change from the ranks they had in the old communicator **comm_old**.

4. int MPI_Comm_rank(MPI_COMM   comm,

                          int   *rank   )

The **MPI_Comm_rank** operation returns the rank of the processor in the communicator **comm**.

5. int MPI_Comm_size(MPI_COMM comm,
                       int   *p   )

The MPI_Comm_size operation returns the number processors in the communicator comm.

6. int MPI_Comm_spawn(char   *command,
                        char   *argv[],
                        int   maxprocs,
                        MPI_Info   info,
                        int   root,
                        MPI_Comm   comm,
                        MPI_Comm   *intercomm,
                        int   errcodes[ ]   )

New processes can be created and connected dynamically in MPI using the MPI_Comm_spawn command. Each new process is mapped to a processor and given a rank. The command parameter is the name of the executable to be spawned and argv are the arguments to it. maxprocs is the maximum number of copies of this executable to make (similar to the "-np" argument of mpirun). Info is a set of key-value pairs with information to the runtime system of the MPI system for starting the process. The first five parameters are interpreted only for the spawning process with rank root. The communicator comm is the group of spawning processes. The communicator intercom allows both spawned and spawning processes to communicate. The array errcodes is contains one code for each spawned process.

7. int MPI_Comm_split (MPI_Comm   comm,
                        int   partition,
                        int   rank,
                        MPI_Comm   *comm_out   )

The operation MPI_Comm_split subdivides the processors in a communicator into disjoint subgroups based on the partition argument.

8. int MPI_Dims_create(int   nnodes,

                            int   ndims,

                            int   *dims   )

MPI_Dims_create creates a Cartesian grid of processors. The nnodes parameter is the total number of processors that will be in the grid. The ndims parameter is the number of dimensions in the grid. The result of the function is an array dims which contains the number of processors in each dimension of the grid. In fact, this argument can be an input to set desired amounts of processors on each dimension. If, however, it is initialized to zero, then the number of processors in each dimension will be all automatically determined to as close to a square grid as possible.

9. int MPI_Finalize()

The MPI environment is terminated by the MPI_Finalize operation; it should be the last MPI function called in a program.

10. int MPI_Gather (void   sendbuf,

                     int   sendcnt,

                     MPI_Datatype   sendtype,

                     void   *recvbuf,

                     int   recvcount,

                     MPI_Datatype   recvtype,

                     int   root,

                     MPI_Comm   comm   )

The gather operation MPI_Gather fills an array on the root processor with equal-sized data blocks sent from each processor. It is a collective communication function — that is, all the processors in the communicator (MPI_COMM_WORLD in general) participate in the operation. The first three parameters give the address of the buffer on the sending processor, sendbuf, its size sendcnt, and its type, sendtype. The second three give the address, recvbuf, size recvcount, and type recvtype, on the receiving (root) processor. The final two are the index of the root processor and the communicator comm.

11. int MPI_Init(　　　int　　　argc,
　　　　　　　　　　　char　　**argv　　)

The MPI_Init operation must be the first MPI call in a program. It initializes the MPI environment.

12. int MPI_Iprobe(　　int　　　　　　source,
　　　　　　　　　　　int　　　　　　tag,
　　　　　　　　　　　MPI_Comm　　comm,
　　　　　　　　　　　int　　　　　　*flag,
　　　　　　　　　　　MPI_Status　*status　)

MPI_Iprobe tests to see if there is a message from source waiting on the port tag and returns the result in flag.

13. int MPI_Recv(　　　void　　　　　　*buf,
　　　　　　　　　　　int　　　　　　count,
　　　　　　　　　　　MPI_Datatype　datatype,
　　　　　　　　　　　int　　　　　　source,
　　　　　　　　　　　int　　　　　　tag,
　　　　　　　　　　　MPI_Comm　　comm,
　　　　　　　　　　　MPI_Status　*status　)

The receive operation MPI_Recv is a point to point message receive operation. The address of the buffer for receiving data is given in buf, its size in count and its type in datatype. The rank of the source processor is in source, and comm is the communicator. The tag parameter allows for some selectivity in receiving messages. The parameter, status, gives information about the source and tag arguments of the received message in the case that the receive operation does not specify specific sender or tag parameters.

14. int MPI_Reduce (　void　　　　　　*sendbuf,
　　　　　　　　　　　void　　　　　　*recvbuf,
　　　　　　　　　　　int　　　　　　count,
　　　　　　　　　　　MPI_Datatype　datatype,
　　　　　　　　　　　MPI_Op　　　op,
　　　　　　　　　　　int　　　　　　root,
　　　　　　　　　　　MPI_Comm　　comm　　　)

The collective communication function MPI_Reduce will perform a *reduction operation* on data from every processor and place the result of the operation on the root processor. A reduction operation is a binary associative operation (these include sum and product, but also maximum, minimum and logical operations; see Appendix III for a full list of MPI's built-in reduction operations). The parameter sendbuf is the address of the subtotal and recvbuf is the address of the final reduced total. If a list (array) of items is being separately reduced, then count can be used to indicate the size of the array. The reduction operation is specified by the value of op.

15. int MPI_Scatter (void   *sendbuf,
           int   sendcnt,
           MPI_Datatype sendtype,
           void   *recvbuf,
           int   recvcnt,
           MPI_Datatype recvtype,
           int   root,
           MPI_Comm comm   )

The scatter operation MPI_Scatter distributes equal-sized data blocks to each processor. It is a collective communication function — that is, all the processors in the communicator (MPI_COMM_WORLD in general) participate in the operation. The sendbuf parameter is the address of the array (buffer) being scattered. The sendcnt parameter contains the size of the block to be sent to each processor and sendtype is the type of data elements in the block.

16. int MPI_Scatterv (void   *sendbuf,
           int   *sendcnts,
           int   *displs,
           MPI_Datatype sendtype,
           void   *recvbuf,
           int   recvcnt,
           MPI_Datatype recvtype,
           int   root,
           MPI_Comm comm)

MPI_Scatterv allows for varying amounts of data to be scattered to each processor. It is a collective communication function — that is, all the processors in the communicator (MPI_COMM_WORLD in general) participate in the operation. As with **MPI_Scatter**, the **sendbuf** parameter is the address of the array (buffer) being scattered and **recvbuf** is the address of the array (buffer) into which the scatter elements are to be placed. The **displs** and **sendcnts** parameters are arrays containing the index and size of the block to be sent to each processor.

17. int MPI_Send(void   *buf,
                        int   count,
                        MPI_Datatype datatype,
                        int   dest,
                        int   tag,
                        MPI_Comm   comm   )

The **MPI_Send** operation is a point to point send operation. The address of the data to be sent is given in **buf**, its size in **count** and its type in **datatype**. The rank of the destination processor is in **dest**, and **comm** is the original, non-grid communicator. The **tag** parameter allows for some selectivity in receiving messages.

Bibliography

1. Agre, P. and Chapman, D., PENGI: An implementation of a theory of activity. In *Proceedings of the Sixth National Conference on Artificial Intelligence (AAAI-87)*, pages 268–272, Seattle WA, 1987.
2. Andersson, L.A.A. and Nygards, J., On multi-robot map fusion by inter-robot observations, *12th International Conference on Information Fusion (FUSION '09)*, Seattle WA, 2009.
3. Arbib, M.A., Iberall, T., and Lyons, D., Coordinated Control Programs for Movements of the Hand, in *Hand Function and the Neocortex*, A.W. Goodwin and I. Darian-Smith (Eds), Berlin, Springer-Verlag (1985).
4. Arbib, M.A., Iberall, T., and Lyons, D.M.,Coordinated Control Programs for Movements of the Hand, *Exp. Brain Res. #10*, Springer-Verlag, 1985
5. Arbib, M.A., *The Metaphorical Brain 2*. Wiley & Sons 1989.
6. Arkin, R.C., *Behavior-Based Robotics*. MIT Press 1998.
7. Asbeck, A.T., Kim, S., Cutkosky, M.R., Provancher, W.R., and Lanzetta, M., Scaling hard vertical surfaces with compliant microspine arrays, *Proceedings, Robotics Science and Systems*, Cambridge MA, 2005.
8. Szatmary, B., Fleischer, J., Hutson, D., Moore, D., Snook, J., Edelman, G. M., and Krichmar, J., A Segway-based human-robot soccer team, *IEEE International Conference on Robotics and Automation*, Orlando FL, 2006.
9. Baase, S. and van Gelder, A., *Computer Algorithms*. Addison-Wesley 1999.
10. Bar-Shalom, Y. and Fortmann, T., *Tracking and Data Association*. Academic Press 1988.

11. Bekey, G.A., *Autonomous Robots.* MIT Press 2005.
12. Besl P. and N. McKay. A Method for Registration of 3–D Shapes, *IEEE Trans. on PAMI*, 14(2):239–256, February 1992.
13. Bilmes, J., A Gentle Tutorial on the EM Algorithm Including Gaussian Mixtures and Baum-Welch, *ICSI Technical Report TR-97-021*, May 1997.
14. Birchfield S.T. and Sriram Rangarajan, Spatial Histograms for Region-Based Tracking, *ETRI Journal*, V29, N5, October 2007.
15. Bookman, C., *Linux Clustering.* New Riders 2003.
16. Borenstein, J., Experimental Results from Internal Odometry Error Correction with the OmniMate Mobile Robot, *IEEE Trans. on Robotics and Automation*, 14(6): pp. 963–969, December 1998.
17. Bradski, G. and Kaehler, A., *Learning OpenCV.* O'Reilly 2008.
18. Braunl, T., Feyrer, S., Rapf, W., and Reinhardt, M., *Parallel Image Processing.* Springer-Verlag, 2001.
19. Brooks, R.A., A robust layered control system for a mobile robot, *IEEE Journal of Robotics and Automation*, 2(1):14–23, 1986.
20. Brummit, B. and Stentz, A., Dynamic Mission Planning for Multiple Mobile Robots, *Proc. IEEE Int. Conf. on Robotics and Automation*, Minneapolis MN, 1996.
21. Chamberlain, B.L., Sung-Eun Choi, E., Christopher Lewis, Calvin Lin, Lawrence Snyder, and Derrick Weathersby, W., ZPL: A machine independent programming language for parallel computers. *IEEE Transactions on Software Engineering*, 6(3):197–211, March 2000.
22. Chapman, B., Jost, G., and van der Pas, R., *Using OpenMP: Portable Shared Memory Parallel Programming.* MIT Press 2008.
23. Cheng, D.Y., A Study of Parallel Programming Languages and Tools, *NASA Ames Research Center Report RND-93-005*, March 1993.
24. Choset, H., Lynch, K., Hutchinson, S., Kantor, G., Burgard, W., Kavraki, L., and Thrun, S., *Principles of Robot Motion.* MIT Press 2005.
25. Cox, I.J. and Hingorani, S.L., An Efficient Implementation and Evaluation of Reid's Multiple Hypothesis Tracking Algorithm for Visual Tracking, *Int. Conf. on Pattern Recog.* (1994) pp. 437–442.
26. Dempster, P., Laird, N.M., and Rubin, D.B., Maximum Likelihood from Incomplete Data via the EM Algorithm, *Journal of Royal Statistics Society*, Vol. B-39, 1977.

27. de Souza, P.S.L., Sabourin, R., de Souza, S.R.S., and Borges, D.L., A low-cost parallel K-means VQ algorithm using cluster computing, *Seventh International Conference on Document Analysis and Recognition (ICDAR'03)*, Edinburgh Scotland, 2003.

28. Dellaert, F., Fox, D., Burgard W., and Thrun, S., Monte Carlo Localization for Mobile Robots, *IEEE International Conference on Robotics and Automation (ICRA99)*, Detroit MI, May 1999.

29. Diosi, A. and Kleeman, L., Laser Scan Matching in Polar Coordinates, *Proc. Int. Robots and Systems (IROS)*, Edmonton Canada, 2005.

30. Dudek, G. and Jenkin, M., *Computational Principles of Mobile Robotics*. Cambridge Press 2000.

31. Elfes, A., Using Occupancy Grids for Mobile Robot Perception and Navigation, *IEEE Computer*, June 1989.

32. Ellore, B.K., Dynamically expanding occupancy grids. *Master's thesis, Texas Tech University*, 2002.

33. Engelberger, J.F., *Robotics in Practice*. MIT Press 1989.

34. Fikes, R.E. and Nilsson, N., STRIPS: A new approach to the application of theorem proving to problem solving. *Artificial Intelligence*, 5(2):189–208, 1971.

35. Firby, J.A., An investigation into reactive planning in complex domains. In *Proceedings of the Tenth International Joint Conference on Artificial Intelligence (IJCAI-87)*, pages 202–206, Milan Italy, 1987.

36. Flynn, M., Some Computer Organizations and Their Effectiveness, *IEEE Transactions on Computing*, Vol. C-21, pp. 948, 1972

37. Fox, D., Ko, J., Konolige, K., Limketkai, B., Schulz, D., and Stewart, B., Distributed Multi-Robot Exploration and Mapping, *Proc. of the IEEE: Special Issue, Multi-Robot Systems* V94 N7, 2006.

38. Goto Y. and Stenz A., The CMU System for Mobile Robot Navigation, *IEEE Int. Conf. on Robotics and Automation*, Raleigh NC, pp. 99–105, 1987.

39. Graham, J.H., Special Computer Architectures for Robotics: Tutorial and Survey, *IEEE Trans. on Robotics and Automation*, 5(5): 1989.

40. Gropp, E., Lusk, E., and Sterling, T., Beowulf Cluster Computing with Linux, *MIT Press* 2003.

41. Henrich, D. and Honiger, T., Parallel Processing Approaches in Robotics, *IEEE International Symposium on Industrial Electronics (ISIE'97)*, Guimarães Portugal, July 7–11, 1997.
42. Hillis, W.D., *The Connection Machine*. MIT Press 1989.
43. Hoare, C.A.R., *Communicating Sequential Processes*. Prentice-Hall 1985.
44. Hough, P., Methods and means for recognizing complex patterns, *US Patent #3,069,654*, 1962.
45. Isard, M. and Blake A., Condensation — Conditional density propagation for visual tracking, *International Journal of Computer Vision* 29:5–28, 1998.
46. Jeannerod, M., Arbib, M.A., Rizzolatti, G., and Sakata, H., Grasping objects: the cortical mechanisms of visuomotor transformation, *Trends Neurosci*, 18:314–320, 1995.
47. Jones, J., *Robot Programming: A Practical guide to Behavior-Based Robotics*. McGraw-Hill 2004.
48. Joshi, M.N., Parallel K-Means Algorithm on Distributed Memory Multiprocessors, *Computer*, April, 2003.
49. Kerridge, J., *Occam Programming: A Practical Approach*. Wiley-Blackwell, 1987.
50. Klechenov, A., Gupta, A.k., Wong, W.F., Ng, T.K., and Leow, W.K., Real-time Mosaic for Multi-Camera Videoconferencing, *MIT-Singapore Alliance Symposium*, Singapore 2003.
51. Kuhn, M., Parallel Image Registration in Distributed Memory Environments, *M.S. Thesis*, Swiss Federal Institute of Technology, 2004.
52. Lakaemper, R., Adluru, N., Latecki, L.J., and Madhavan, R., Multi Robot Mapping using Force Field Simulation. *Journal of Field Robotics*, 24(8/9):747–762, 2007.
53. Bongo, L., Anshus, O., and Bjørndalen, J., Evaluating the Performance of the Allreduce Collective Operation on Clusters: Approach and Results, *Technical Report 2004–48*, University of Tromso, 2004.
54. Lewis, M.A., Fagg, A.H., and Bekey, G.A., The USC Autonomous Flying Vehicle: an Experiment in Real-Time Behavior-Based Control, *Proceedings of the IEEE Conference on Robotics and Automation*, May 1993, Atlanta Georgia.

55. López de Teruel, P., García, J., and Acacio, M., The Parallel EM Algorithm and its Applications in Computer Vision, *Proceedings Parallel and Distributed Processing Techniques and Applications*, Las Vegas Nevada, 1999.

56. Lowe, D., Distinctive image features from scale-invariant keypoints, *International Journal of Computer Vision*, 60(2):91–110, 2004.

57. Lucke, R.W., *Building Clustered Linux Systems*. Prentice Hall 2005.

58. Lyons, D., Selection and Recognition of Landmarks Using Terrain Spatiograms, *IEEE/RSJ International Conference on Intelligent RObots and Systems (IROS)*, Tapei, Taiwan, October 2010.

59. Lyons, D. and Hsu, D.F., Rank-based Multisensory Fusion in Multitarget Video Tracking, *IEEE Intr. Conf. on Advanced Video & Signal-Based Surveillance*. Como Italy, 2005.

60. Lyons, D.M. and Arbib, M.A., A Formal Model of Computation for Sensory-based Robotics, *IEEE Transactions on Robotics and Automation* 5(3), June 1989.

61. Lyons, D.M., Sharing Landmark Information using Mixture of Gaussian Terrain Spatiograms, *IEEE/RSJ International Conference on Intelligent RObots and Systems (IROS)*, St. Louis MO, October 2009.

62. Lyons, D.M. and Arkin, R., Towards Performance Guarantees for Emergent Behavior, *IEEE International Conference on Robotics and Automation*, New Orleans LA, 2004.

63. Lyons, D.M., Detection and Filtering of Landmark Occlusions using Terrain Spatiograms, *IEEE Int. Conference on Robotics and Automation*, Anchorage Alaska, May 2010.

64. Lyons, D.M. and Isner, G.R., Evaluation of a Parallel Algorithm and Architecture for Mapping and Localization, *7th International Symposium on Computational Intelligence In Robotics and Automation, CIRA 2007*, Jacksonville FL, June 20–23, 2007.

65. Lyons, D.M., Representing and Analyzing Action Plans as Networks of Concurrent Processes, *IEEE Transactions on Robotics and Automation* 9(3), June 1993.

66. McGraw, J., Skedzielewski, S., Allan, S., Grit, D., Oldehoeft, R., Glauert, J., Dobes, I., and Hohensee, P., SISAL-Streams and Iterations in a Single Assignment Language, Language Reference Manual, version 1. 2. *Technical Report TR M-146, University of California — Lawrence Livermore Laboratory*, March 1985.

67. Medeiros, H., Gao, X.T., Kleihorst, R.P., Park, J., and Kak, A.C., A parallel color-based particle filter for object tracking, *The 6th ACM Conference on Embedded Networked Sensor Systems*, Raleigh NC, November 5–November 7, 2008.
68. Message Passing Interface Forum, MPI: A Message-Passing Interface Standard, Version 2.2, 2009.
69. Mobilerobots Inc. Pioneer 3 & Pioneer 2 H8-Series Operations Manual, 2003.
70. Montemerlo, M., *et al.*, Junior: The Stanford Entry in the Urban Challenge, *Journal of Field Robotics*, 25(9), September 2008.
71. Moravec, H. and Elfes, A., High Resolution Maps from Wide Angle Sonar, *IEEE International Conference on Robotics and Automation*, Washington DC, 1985.
72. Mundhenk, N., *et al.*, Low cost, high performance robot design utilizing off-the-shelf parts and the Beowulf concept, The Beobot project, *Proceedings, SPIE Intelligent robots and computer vision XXI: algorithms, techniques, and active vision*, Providence RI, 28–29 October 2003.
73. Nof, S.Y., *Handbook of Industrial Robotics*, Wiley 1999.
74. Nuchter, A., Parallelization of Scan Matching for Robotic 3D Mapping, *Proc. 3rd European Conf. on Mobile Robots*, 2007.
75. Oliveira, P. and du Buf, H., SPMD Image Processing on Beowulf Clusters: Directives and Libraries, *Proceedings of the 17th International Symposium on Parallel and Distributed Processing*, 2003.
76. Ouerhani, N., von Wartburg, R., Hugli, H., and Muri, R., Empirical Validation of the Saliency-based Model of Visual Attention, *Electronic Letters on Computer Vision and Image Analysis*, 3(1):13–24, 2004.
77. Parker, J., *Algorithms for Image Processing and Computer Vision*, Wiley 1997.
78. Pfister, G., *In Search of Clusters*. 2nd *Ed.*, Prentice Hall, 1998.
79. Patarasuk P. and Xin Yuan, Bandwidth optimal all-reduce algorithms for clusters of workstations, *Journal of Parallel and Distributed Computing*, 69(2), February 2009.
80. Pjesivac-Grbovic, J., Angskun, T., Bosilca, G., Fagg, G. E., Gabriel, E., and Dongarra, J.J., Performance analysis of MPI collective operations, *Cluster Computing*, 10(2):127–143, 2007.

81. Pomerleau, D., Neural Networks for Intelligent Vehicles, *Proceedings of the Intelligent Vehicles Conference*, Tokyo, Japan 1993.
82. Quinn, M.J., *Parallel Programming in C with MPI and OpenMP*, McGraw-Hill 2004.
83. Rao, B., *Data Association Methods for Tracking Systems*, in *Active Vision*, A. Blake, Yuille, A., Editors. 1992, MIT Press. pp. 91–106.
84. Rasmussen, C. and Hager, G., Joint Probabilistic Techniques for Tracking Multi-Part Objects. *Proc. Computer Vision & Pattern Recognition*, Santa Barbara CA, pp. 16–21, 1998.
85. Rosselot, D. and Hall, E., Processing real-time stereo video for an autonomous robot using disparity maps and sensor fusion, *SPIE Intelligent Robots and Computer Vision XXII: Algorithms, Techniques, and Active Vision*, pp. 70–78, 2004.
86. Sack, D. and Burgard, W., A comparison of methods for line extraction from range data, 5th *IFAC Symposium on Intelligent Autonomous Vehicles*, Lisbon, Portugal, 2004.
87. Se, S., David Lowe, Jim Little, Local and Global Localization for Mobile Robots using Visual Landmarks, *Proceedings of the International Conference on Intelligent Robots and Systems* (IROS), pp. 414–420, 2001.
88. Sermanet, P., Hadsell, R., Scoffier, M., Grimes, M., Ben, J., Erkan, A., Crudele, C., Muller, U., and LeCun, Y., A Multi-Range Architecture for Collision-Free Off-Road Robot Navigation, *Journal of Field Robotics*, 26(1):58–87, January 2009.
89. Sloan, J., *High performance Linux Clusters*, O'Reilly 2005.
90. Squyres, J., Lumsdaine, A., and Stevenson, R., A toolkit for parallel image processing, *Proceedings of the SPIE Conference on Parallel and Distributed Methods for Image Processing*, 1998.
91. Steenstrup, M., Arbib, M.A., and Manes, E.G., Port automata and the algebra of concurrent process, *J. Computer System Sciences*, 27(1):29–50, 1983.
92. Thrun, S., Burgard, W., and Fox, D., A Probabilistic Approach to Concurrent Mapping and Localization for Mobile Robots, *Machine Learning and Aut. Robots*, 31/5:1–25, 1998.

93. Thrun, S., Fox, D., Burgard, W., and Dellaert, F., Robust monte carlo localization for mobile robots, *Artificial Intelligence 128* (1–2):99–141, 2001.
94. Thrun. S., *Robotic mapping: A survey*. In G. Lakemeyer and B. Nebel, editors, *Exploring Artificial Intelligence in the New Millennium*. Morgan Kaufmann, 2002.
95. Usher, K., Obstacle avoidance for a non-holonomic vehicle using occupancy grids, *Proceedings of the 2006 Australasian Conference on Robotics & Automation*, Auckland, New Zealand, December 2006.
96. Venkataraman S.T. and. Iberall, T., (Eds.) Dexterous Robot Hands. *New York: Springer Verlag*, 1990.
97. Wilkes, D., Dudek, G., Jenkin, M., and Milios, E., Modeling Sonar Range Sensors, in: (Archibald and Petriu, Eds) *Advances in Machine Vision: Strategies and Applications*, World Scientific Press, Singapore 1992.
98. Wilkinson, B. and Alen, M., *Parallel Programming*. 2nd Ed. Prentice-Hall 2005.
99. Yuh, J., Design and Control of Autonomous Underwater Robots: A Survey, *Autonomous Robots*, 8(1), January 2000.

Index